Sustainable Rural Development and Climate Change

Sustainable Rural Development and Climate Change

Satish Chandra

AMIGA PRESS. INC
Delhi -110051

AMIGA PRESS. INC

H.O :- 28-A, Indra Park,

Chander Nagar, Delhi :- 110051 (India)

Tel :- 011-22504028, +91-98736-57000

Email :- amigapress@gmail.com
amigaeditorial@gmail.com
amigapub@gmail.com

Website :- www.amigapress@co.in

This edition is published by Amiga Press Inc.

ISBN : 978-93-87295-46-9

First Published 2019

Printed and bound by Replika Press Pvt. Ltd.

Preface

Sustainable development is widely advocated as the way to deal with the issues that fill the pages of this publication. The concept is, however, hard to apply in practice, and easy to manipulate. These days many promote 'sustainability', including some who are only interested in sustaining that which benefits them directly. But to make the world a better place for all, current practices must change. The term 'sustainable development' implies that some forms of development cannot be sustained (continued indefinitely). Here we look at why we need a different kind of development, what sustainable development is about, and how the concept can be applied.

Advocates of sustainable development, as we have noted, recognize the social component of development as an essential part of the new paradigm. In doing so, they are validating the importance of a much older perspective. A "human development" approach emphasizing issues of basic needs and equity is well grounded in the history of economic theory. Sudhir Andand and Amartya Sen point out that concerns for these dimensions of economic development start with the earliest economic theorists, and contrast the human development approach to the wealth maximization approach which has dominated modern economics.

Sustainable development has a universal appreciation. At first sight, this is highly positive, as this could signal the entering of a holistic and responsible thinking into the world of politics and society. But as it often happens with other catch phrases that suddenly come into vogue, it fails to be translated into practice, this all the more so because sustainable development can be given several different interpretations.

Climate changes in response to changes in the global energy balance. On the broadest scale, the rate at which energy is received from the sun and the rate at which it is lost to space determine the equilibrium temperature and climate of Earth. This energy is then distributed around the globe by winds, ocean currents, and other mechanisms to affect the climates of different regions. Factors that can shape climate are called climate forcings or "forcing mechanisms". These include such processes as variations in solar radiation, deviations in the Earth's orbit, mountain-building and continental drift, and changes in greenhouse gas concentrations. There are a variety of climate change feedbacks that can either amplify or diminish the initial forcing. Some parts of the climate system, such as the oceans and ice caps, respond slowly in reaction to climate forcings, while others respond more quickly. Forcing mechanisms can be either "internal" or "external". Internal forcing mechanisms are natural processes within the climate system itself (e.g., the meridional overturning circulation). External forcing mechanisms can be either natural (e.g., changes in solar output) or anthropogenic (e.g., increased emissions of greenhouse gases).

This is a reference book. All the matter is just compiled and edited in nature, taken from the various sources which are in public domain.

This book provides deep insight to various dimensions of issues relating to the subject.

—*Editor*

Contents

1

Introduction

Much of research into development issues represents the second basic view, namely that poverty is a development problem where the best solutions are shaped through local participation and policy measures are adapted to local conditions. In other words, this research takes a bottom-up approach to the problem, even though some of the basic causes of poverty can be global.

In terms of environmental research, there has been an increasing focus on the interrelationship between environment and development, and how nature and society affect each other through a complex interplay. This approach also implies that global environmental changes can have very different impacts on different local communities. The IPCC's suggestion that poverty plays an important role in determining how climate changes will affect various populations has two important consequences for policy design principles:

i) That those who are most vulnerable to the impacts of climate change should be the primary target of mitigation policy; poverty issues should be taken into account as a central element in climate measures.

ii) Consequence is that policy design must take into account that the impacts of climate changes and possible effective measures to reduce vulnerability vary both from place to place and from time to time because social conditions change.

It will be just as important to identify how adaptability can be improved among various social groups, as it will be to identify physical climate impacts. Measures should therefore be prepared in cooperation with the local population, where their priorities and conception of the problems become the center of focus.

COMPLIANCE COST ANALYSIS

Compliance cost analyses are undertaken in a number of countries where these are focused on providing an indication of the impacts which a proposed regulation will have on industry and their significance to the competitiveness of the sectors in question. They differ from CBAs in that they focus on estimating the financial costs to different industrial/business sectors, rather than examining costs in economic resource or opportunity cost terms (as required in a CBA). Often such analyses also require that particular attention is paid to the impacts of a new regulation on small companies, given that they are less likely to be able to absorb additional costs. Formal Compliance Cost Assessments (CCAs) are required of all new regulatory proposals within the UK.The guidelines on undertaking a CCA state that the analyses are intended to:

- Show the likely cost implications to business of complying with new or amended regulations;
- Inform Ministers, Members of Parliament, business and other interested parties of the cost implications of regulatory proposals well before a decision is taken to go ahead; and
- Ensure that new or amended regulations do not impose unnecessary burdens on business.

It is recognised that the CCA is not equivalent to a full economic CBA, and indeed the guidelines recognise that full CBAs may be required for those policies giving rise to significant costs. The compliance costs considered within CCAs include all incremental recurring (annual) and non-recurring (capital and other one-off) costs arising from a regulation. Other information required within these analyses includes details of the sectors and number of companies within them likely to be affected, the impacts on small businesses, and the impacts on the UK's competitive position.

Environmental Impact Assessment

Environmental Impact Assessments (EIAs) are generally thought of as operating at a project level, with specific regulations (such as EEC/85/337) requiring that all major development projects falling under certain headings be subject to one. However, Australia has developed a framework for environmental and health impact assessment (E&HIA) which builds upon the principles of EIA; although it is focused towards project level assessments, it is also applicable at programme, plan and policy levels. This framework sets out a seven step approach involving:

1) Screening: is an E&HIA required?
2) Scoping: what issues must be addressed?

3) Profiling: what is the current status of the affected population or local environment?
4) Risk assessment: what are the risks or benefits?
5) Risk management: can risks be avoided? Are there better alternatives? Can the benefits and risks be priced? etc.
6) Implementation and decision-making: does the assessment provide information for decision-making? Is there a conflict to be resolved? etc
7) Monitoring, environmental and health auditing, post-project evaluation.

The framework bears many similarities to the types of framework which most countries indicate surround their use of CBA or CEA. Perhaps the greatest difference between CBA/CEA on the one hand and EIA on the other is the additional focus given by the latter to consultation and stakeholder involvement. Although this is not reflected in the above seven steps, consultation and stakeholder involvement tend to be a more formal requirement within EIA procedures. Another potential aspect of EIA, which varies from the use of CBA or CEA, is the explicit requirement of monitoring and post-project evaluation. Although some countries and some government departments will have requirements for such ex-post evaluations of policy, these will not necessarily form an automatic step within the appraisal process.

THE ENVIRONMENT IN INDIA

In the subcontinent of India, a large part of humanity is huddled together. The second most populated country in the world, with 635 million people, increasing at the rate of 2 percent a year, it will have approximately 900 million people at the end of the century. Seventy- eight percent of the people live in 560,000 villages and the rest in 4,000 cities and towns. The per capita income of India is approximately $150 per year. India is expected to produce 130 million metric tons of food this year. For centuries a vast majority of Indians have depended on wood and cow dung for energy. Vast areas of India have been deforested through the centuries and there is no abeyance of this process in sight. Simultaneously, with deforestation there has been a continuous danger to the wildlife. Recently, a survey identified that approxi-mately 52 animal species in India are on the endangered list.

More seriously, however, India faces a crisis of clean and potable water supply. In the country as a whole, approximately 60 percent of the people do not get safe water for drinking and, personal hygiene. Recently, a Union Minister said that over 25 percent of the urban population in India does not have safe water and 65 percent do not have a sewage system. It is estimated that the daily average supply of water per person in Delhi, Calcutta, Bombay, and Madras is 180, 140, 130, and 80 liters, respectively. Limited supply, poor quality of water, and waste are some of the major problems affecting urban areas.

In the rural areas the water situation is disgraceful despite the fact that under the Fifth Five-Year Plan it was expected that 90 percent of all villages would be provided with safe drinking water at a cost of Rs 550 crores. * Some time ago, the Central Water and Power Commission said that it would cost India Rs 5800 crores to provide minimum water supply to its rural population and an optimum level of water supply and sewerage facilities to the urban population. At the present prices, the figures will be no less than Rs 15,000 crores, and possibly even higher. Considering the shortage of re sources, it is fairly certain that the water crisis in India will deepen each year. A continuous supply of fresh water for everyone, as is avail able in industrialized countries, will remain an illusion for India far into the twenty-first cent * dB = decibels.

POLLUTION IN INDIA

Rachel L. Carson once said,

"For the first time in the history of the world, every human being is now subjected to contact with dangerous chemicals, from the moment of conception until death."

Despite the fact that India is a rural country, the endless streams of effluents and poisonous gases discharged by industries, power generators, transportation systems, and other human activities pollute its air, water, and land. Even modern agriculture has aggravated the problem of pollution by supplying nitrogen fertilizers, insecticides, pesticides and herbicides, which are washed off by rains into rivers and the sea. More seriously, many heavy metals, such as mercury, lead, and cadmium, are found in the waters of the River Ganges and the Arabian Ocean near Bombay. A recent survey of air pollution by the Bombay Municipal Corporation has identified heart and lung diseases, such as asthma, bronchitis, emphysema, and even tuberculosis and cancer, related to the air pollution. Even though it is generally recognized that the urban communities suffer from acute forms of pollution, particularly air pollution and noise pollution, it is quite common to find paper, pulp, and 'sugar factories polluting rivers and streams.

Many chronic diseases such as jaundice, cholera, and gastroenteritis, which are related to the impurities of water, are found in villages and cities all over India. The total social and economic cost of pollution is yet to be measured. Except in the major cities, there is no awareness about the environment or pollution. There is no significant movement to control pollution or improve the environment. On the contrary, many planners, administrators, and policymakers live in an euphoria of arrogance and complacency, assuming that everything is fine with the environment and pollution is either impersonal or an inexorable price of technological progress. Sometimes, it is even charged that pollution is a Western technique for keeping a developing country undeveloped. For some people "poverty is the worst pollution." There is no perceptible conservation movement.

Air Pollution

Air and water pollution are the most serious and obvious. By and large, the crisis of air pollution is concentrated in industrial areas. The major air pollutants identified in the cities of India are carbon dioxide, carbon monoxide, particulates, hydrocarbons, sulphur dioxide, and nitrogen oxides, The major sources of air pollution in the urban and industrial areas are industrial processes (40%), transportation (25%), fuel production (25%), others (10%). Industrial processes include cement manufacturing, iron foundries, iron and steel mills, pulp and paper mills, petroleum refineries, phosphate fertilizer plants, coal cleaning, and the smelters of aluminum, copper, lead and zinc. In the city of Bombay alone, air pollution has increased tenfold during the last 25 years. There are approximately 15,000 industries located in the city and 250,000 vehicles. In the winter months, smog is a major menace in Bombay, Calcutta, Baroda, Durgapur, Ahmedabad, and Poona. In all these cities, CO and SO have been significantly in creasing in recent years. The trend extrapolation of the air pollution level in four major metropolitan cities of India has shown that it will be four times greater than what it is today, unless some drastic measures are taken immediately. Unfortunately, there is no Air Pollution Act in India to date. It is hoped that the Act before the Parliament will be passed soon and the necessary measures taken to control air pollution. It would be helpful if the tree plantation programs in the cities were undertaken soon to aid in cleaning the environment.

Water Pollution

Water pollution is equally insidious arid more widespread than air pollution in India. As pointed out earlier, large and small water bodies around the country have become cesspools because of large quantities of effluents discharged indiscriminately. The major industrial offenders are petroleum, chemicals, paper and pulp,and sugar. Another major contributor to water pollution is sewage from cities. Only seven percent of the population in India is covered by drainage systems and only an insignificant amount of sewage is treated.

Development of sewage and drainage facilities has been miserable in most of the urban areas. Out of 4,000 cities and towns, hardly 250 have any sort of sewage system. More than 50 percent of the population in urban areas does not have this basic amenity. It is also well known that most of the coastal areas around Bombay, Calcutta, and other cities are highly polluted. The Fishermen's Association of Bombay pointed out in 1978 that their fish catch had been reduced to hardly 15 percent of what they caught 25 years ago, and most of the time they have to go far out into the sea to get any significant catch. Most of the beaches around Bombay, Calcutta, and Madras are not fit for swimming.

In 1974, the government of India passed the Water Pollution Act, requiring that each state set up water pollution control authorities. Most of these authorities lack funds, trained people, and facilities to implement the Act. Yet, within the last four years significant steps were taken to control water pollution problems in major cities by requiring the industries to set up pollution control measures and treat their effluent. The standards of effluent control are high, and it will require at least five more years before a significant impact on the quality of water will be perceptible. The lack of sewerage and drainage facilities is a serious urban problem. It is estimated that the cost of cleaning rivers and coastal areas of India will be no less than Rs 7,000 crores. There are neither plans nor finances for such a gigantic task. Moreover, it is not realistic to expect a large number of cities and towns to have sewerage facilities when their basic problem is to supply the minimum fresh water.

However, in a warm country like India, many economical and simple methods of sewage treatment-such as oxidation ponds-cap be de signed. In the city of Bombay, plans are underway to dispose of sewage through outfalls, 10 kilometers into the sea. In Ahmedabad, where there is a sewage treatment plant, it is not used. Instead, because of its high quality of nutrients, sewage is utilized for growing grass. Some industries, like Union Carbide, re cycle their effluent. In short, extensive efforts are underway to protect the water bodies from pollution. Like supplying potable water, wastewater treatment is an expensive proposition. With increasing industrialization and large concentration of population in urban areas, not only the need for a water supply has been accelerated, but also the amount of sewage produced has proportionately increased. Rivers and oceans have become receptacles for enormous amounts of industrial waste. It is seldom realized that the water system of the earth is rather fragile and its capacity to recycle and supply fresh water perennially is limited. In fact, several types of effluents have already destroyed marine life and depleted oxygen supply. Some of the poisonous elements even reach people through food chains, thus causing damage to our health. Oceans and rivers cannot be a panacea for liquid and solid waste problems. A scientific system of treating industrial waste is probably the only answer. Extensive re search in this area will open new vistas for the development and protection of water resources.

Noise Pollution

Among the various enemies of man, one that is most insidious, disquieting, and dangerous is noise, which menaces a large part of humanity now living as members of industrial urban civilization. Each year an increasing number of people are becoming urbanized; the number of cities is increasing, and their sizes are enlarging. In each city, man is exposed to a larger number of technologies, their products and by products. One poisonous by-product of these technologies is pandemonium, which is becoming pain fully louder each year, and may drive cacophonous urban dwellers to permanent silence. In India the situation is dismally pathetic. Some years ago the National Physical Laboratory conducted studies on traffic noise in major cities of India. The studies revealed that the street noise during the day in cities like Bombay and Delhi was intolerably as high as 90 dB* and it was seldom below 60 dB. Similar conditions exist in many other cities and towns around the country. In a recent survey by the Urban Development Institute, it was observed that Bombay was the noisiest city in the country with the largest number of moving vehicles (250,000). Its average daytime noise level was 75 dB, the highest being 105 dB near the airport. The main island continues to remain noisy even during the night, whereas the noise level in the suburbs begins to subside after 11 p.m. The main city can be called the "Island of Cacophony."

If the noise trends in Bombay continue at the present rate we shall, overall, add one decibel a year and reach a dangerous point by the turn of the century. Of the 7.6 million people living in Bombay, at least 30 percent are affected adversely by noise. Besides hearing impairment, they suffer from various other health hazards, including many minor disruptions such as sleeplessness, communication disturbances, and other ills. The industrial aspects of the life we live create much of the noise, but a great deal of it is the result of improper values and unconcern for the environment.

Unlike the problems of other environmental pollutants, there is sufficient knowledge today to control and abate most sources of noise. Various technological, socio-cultural, educational, and legislative * A crore is 10,000,000; thus Rs 550 crore is 5,500,000,000 rupees. A rupee is about $12; there are approximately 8.25 rupees to the American dollar. Measures are required to build a peaceful environment. There is no Noise Pollution Act in India today, yet controlling noise pollution is cheaper than controlling any other pollution.

Land Pollution

The generation of solid waste in a developing country like India is quite limited. An average Indian produces 1 pound of garbage per day, compared to an average European who produces 4 pounds per day and an American who produces five pounds per day. Fortunately, urban refuse in the cities .of India is mostly organic, containing very little plastic. Approximately 10,000 people earn their living by collecting papers, metals, and bottles, earning Rs 5-30 a day. The removal of recyclable materials is common all over India. In this sense, India is a highly conserving country. Unfortunately, refilling the land has wasted the remaining solid waste. Recognizing that urban solid waste has a high content of organic materials, only recently a 300-ton plant has been set up in Bombay to compost it. Even though such compost would be 25 percent more expensive than the chemical fertilizers, its value as soil conditioner has been well recognized; and it is hoped that soon many such plants will be set up in major cities of India to use solid waste economically. In the rural areas, villagers always work in the closed-loop system of utilizing their limited solid waste for the purpose of agriculture or energy. During the last 10 years, extensive efforts have been underway to set up gobar gas plants to produce methane gas and use the remaining waste as fertilizer. Already approximately 80,000 such plants have been set up. Even though there is no Act to control solid waste pollution, this problem can be tackled more easily, with out sophisticated technologies, for generating resources.

CLIMATE CHANGE IMPACT ON AGRICULTURE

Despite technological advances, such as improved varieties, genetically modified organisms, and irrigation systems, weather is still a key factor in agricultural productivity, as well as soil properties and natural communities. The effect of climate on agriculture is related to variabilities in local climates rather than in global climate patterns. The Earth's average surface temperature has increased by 1 degree F in just over the last century. Consequently, agronomists consider any assessment has to be individually consider each local area.

On the other hand, agricultural trade has grown in recent years, and now provides significant amounts of food, on a national level to major importing countries, as well as comfortable income to exporting ones. The international aspect of trade and security in terms of food implies the need to also consider the effects of climate change on a global scale. A study published in *Science* suggests that, due to climate change, "southern Africa could lose more than 30% of its main crop, maize, by 2030. In South Asia losses of many regional staples, such as rice, millet and maize could top 10%".

The 2001 IPCC Third Assessment Report concluded that the poorest countries would be hardest hit, with reductions in crop yields in most tropical and sub-tropical regions due to decreased water availability, and new or changed insect pest incidence. In Africa and Latin America many rainfed crops are near their maximum temperature tolerance, so that yields are likely to fall sharply for even small climate changes; falls in agricultural productivity of up to 30% over the 21st century are projected. Marine life and the fishing industry will also be severely affected in some places.

Climate change induced by increasing greenhouse gases is likely to affect crops differently from region to region. For example, average crop yield is expected to drop down to 50% in Pakistan according to the UKMO scenario whereas corn production in Europe is expected to grow up to 25% in optimum hydrologic conditions.

More favourable effects on yield tend to depend to a large extent on realization of the potentially beneficial effects of carbon dioxide on crop growth and increase of efficiency in water use. Decrease in potential yields is likely to be caused by shortening of the growing period, decrease in water availability and poor vernalization. In the long run, the climatic change could affect agriculture in several ways :

- *productivity*, in terms of quantity and quality of crops
- *agricultural practices*, through changes of water use (irrigation) and agricultural inputs such as herbicides, insecticides and fertilizers
- *environmental effects*, in particular in relation of frequency and intensity of soil drainage (leading to nitrogen leaching), soil erosion, reduction of crop diversity
- *rural space*, through the loss and gain of cultivated lands, land speculation, land renunciation, and hydraulic amenities.
- *adaptation*, organisms may become more or less competitive, as well as humans may develop urgency to develop more competitive organisms, such as flood resistant or salt resistant varieties of rice.

They are large uncertainties to uncover, particularly because there is lack of information on many specific local regions, and include the uncertainties on magnitude of climate change, the effects of technological changes on productivity, global food demands, and the numerous possibilities of adaptation.

Most agronomists believe that agricultural production will be mostly affected by the severity and pace of climate change, not so much by gradual trends in climate.

If change is gradual, there may be enough time for biota adjustment. Rapid climate change, however, could harm agriculture in many countries, especially those that are already suffering from rather poor soil and climate conditions, because there is less time for optimum natural selection and adaption.

Projections

Schneider *et al.*. (2007:787) assessed the literature on key vulnerabilities to climate change. With low to medium confidence, they concluded that for about a 1 to 3°C global mean temperature increase (by 2100, relative to the 1990-2000 average level) there would be productivity decreases for some cereals in low latitudes, and productivity increases in high latitudes. With medium confidence, global production potential was predicted to:

- increase up to around 3°C,
- very likely decrease above about 3 to 4°C.

Most of the studies on global agriculture assessed by Schneider *et al.*. (2007:790) had not incorporated a number of critical factors, including changes in extreme events, or the spread of pests and diseases.

Studies had also not considered the development of specific practices or technologies to aid adaptation.

Regional

- Africa:
    - o Africa's geography makes it particularly vulnerable to climate change, and seventy per cent of the population rely on rain-fed agriculture for their livelihoods. Tanzania's official report on climate change suggests that the areas that usually get two rainfalls in the year will probably get more, and those that get only one rainy season will get far less. The net result is expected to be that 33% less maize—the country's staple crop—will be grown. Alongside other factors, regional climate change-in particular, reduced

precipitation-is thought to have contributed to the conflict in Darfur. The combination of decades of drought, desertification and overpopulation are among the causes of the conflict, because the Baggara Arab nomads searching for water have to take their livestock further south, to land mainly occupied by farming peoples.

  - o With high confidence, IPCC (2007:13) concluded that climate variability and change would severely compromise agricultural production and access to food.

- Asia: With medium confidence, IPCC (2007:13) projected that by the mid-21st century, in East and Southeast Asia, crop yields could increase up to 20%, while in Central and South Asia, yields could decrease by up to 30%. Taken together, the risk of hunger was projected to remain very high in several developing countries. More detailed analysis of rice yields by the International Rice Research Institute forecast 20% reduction in yields over the region per degree Celsius of temperature rise. Rice becomes sterile if exposed to temperatures above 35 degrees for more than one hour during flowering and consequently produces no grain.
- Australia and New Zealand: Hennessy *et al.*. (2007:509) assessed the literature for this region. They concluded that without further adaptation to climate change, projected impacts would likely be substantial: By 2030, production from agriculture and forestry was projected to decline over much of southern and eastern Australia, and over parts of eastern New Zealand; In New Zealand, initial benefits were projected close to major rivers and in western and southern areas. Hennessy *et al.*. (2007:509) placed high confidence in these projections.
- Europe: With high confidence, IPCC (2007:14) projected that in Southern Europe, climate change would reduce crop productivity. In Central and Eastern Europe, forest productivity was expected to decline. In Northern Europe, the initial effect of climate change was projected to increase crop yields.
- Latin America: With high confidence, IPCC (2007:14) projected that in drier areas of Latin America, productivity of some important crops would decrease and livestock productivity decline, with adverse consequences for food security. In temperate zones, soybean yields were projected to increase.
- North America:
  - o According to a paper by Deschenes and Greenstone (2006), predicted increases in temperature and precipitation will have virtually no effect on the most important crops in the US.

- o With high confidence, IPCC (2007:14-15) projected that over the first few decades of this century, moderate climate change would increase aggregate yields of rain-fed agriculture by 5–20%, but with important variability among regions. Major challenges were projected for crops that are near the warm end of their suitable range or which depend on highly utilized water resources.

- Polar regions (Arctic and Antarctic):
  - o For the *Guardian* newspaper, Brown (2005) reported on how climate change had affected agriculture in Iceland. Rising temperatures had made the widespread sowing of barley possible, which had been untenable twenty years ago. Some of the warming was due to a local (possibly temporary) effect via ocean currents from the Caribbean, which had also affected fish stocks.
  - o Anisimov *et al.*. (2007:655) assessed the literature for this region. With medium confidence, they concluded that the benefits of a less severe climate were dependent on local conditions. One of these benefits was judged to be increased agricultural and forestry opportunities.
- Small islands: In a literature assessment, Mimura *et al.*. (2007:689) concluded, with high confidence, that subsistence and commercial agriculture would very likely be adversely affected by climate change.

Shortage in Grain Production

Between 1996 and 2003, grain production has stabilized slightly over 1800 millions of tons. In 2000, 2001, 2002 and 2003, grain stocks have been dropping, resulting in a global grain harvest that was short of consumption by 93 millions of tons in 2003. The Earth's average temperature has been rising since the late 1970s, with nine of the 10 warmest years on record occurring since 1995. In 2002, India and the United States suffered sharp harvest reductions because of record temperatures and drought. In 2003 Europe suffered very low rainfall throughout spring and summer, and a record level of heat damaged most crops from the United Kingdom and France in the Western Europe through Ukraine in the East. Bread prices have been rising in several countries in the region.

Poverty Impacts

Researchers at the Overseas Development Institute (ODI) have investigated the potential impacts climate change could have on agriculture, and how this would affect attempts at alleviating poverty in the developing world. They argued that the effects from moderate climate change are likely to be mixed for developing countries. However, the vulnerability of the poor in developing countries to short term impacts from climate change, notably the increased frequency and severity of adverse weather events is likely to have a negative impact. This, they say, should be taken into account when defining agricultural policy.

Crop Development Models

Models for climate behaviour are frequently inconclusive. In order to further study effects of global warming on agriculture, other types of models, such as *crop development models, yield prediction,* quantities of *water or fertilizer consumed,* can be used. Such models condense the knowledge accumulated of the climate, soil, and effects observed of the results of various agricultural practices. They thus could make it possible to test strategies of adaptation to modifications of the environment.

Because these models are necessarily simplifying natural conditions (often based on the assumption that weeds, disease and insect pests are controlled), it is not clear whether the results they give will have an *in-field* reality. However, some results are partly validated with an increasing number of experimental results. Other models, such as *insect and disease development* models based on climate projections are also used (for example simulation of aphid reproduction or septoria (cereal fungal disease) development). Scenarios are used in order to estimate climate changes effects on crop development and yield. Each scenario is defined as a set of meteorological variables, based on generally accepted projections. For example, many models are running simulations based on doubled carbon dioxide projections, temperatures raise ranging from 1°C up to 5°C, and with rainfall levels an increase or decrease of 20%. Other parameters may include humidity, wind, and solar activity. Scenarios of crop models are testing farm-level adaptation, such as sowing date shift, climate adapted species (vernalisation need, heat and cold resistance), irrigation and fertilizer adaptation, resistance to disease. Most developed models are about wheat, maize, rice and soybean.

LOCAL INTERACTIONS BETWEEN VARIOUS ENVIRONMENTAL PROBLEMS

An important aspect of local adaptability to environmental change is the interaction between various environmental problems at a local level. To an impoverished farmer, the degradation of several natural resources that together form his or her subsistence base is considered a single problem, even though the causes of the degradation are considered to be separate phenomena at a global level. Environmental problems that are addressed separately at an international level - such as climate, biodiversity, and desertification - must at be tackled holistically at the local level. There are several practical examples of the interaction between several "different" environmental issues in locally adapted measures. One study in two arid agricultural areas (one in the Kitui district in Kenya and the other in the Same district in Tanzania) shows that biodiversity in the form of local tree and plant species distributed throughout the farms represents an alternative income source for poor farmers when crops fail. For example, local wood is used to make stools, kitchen equipment, chicken coops, and so on. These products are sold on the local markets. Lumber is also used to burn charcoal, which is sold to urban areas and cities. Leaves and seeds from certain trees are used as feed for goats and cattle. Drought resistant indigenous fruits are an important source of nutrition for both children and adults when there is little food. Thus preservation of local knowledge and biodiversity in cultivated areas helps enhance adaptability and reduce vulnerability to extreme climate events such as drought and flooding (Eriksen 2000). Preservation of the natural vegetation is also useful in combating desertification. Moreover, particular ridges and hills covered with natural vegetation in dry areas have great value with respect to biodiversity. These areas often have different compositions of plant and animal species that otherwise are not found in the surrounding lowlands (Gachathi 1995). These ridges, which have higher precipitation than the lowlands, form important parts of local watersheds. In addition, there is a better microclimate (local temperatures and moisture) around these forested areas, particularly in arid regions that are dependent on the forests as water sources. Where ridge vegetation is well preserved, forest streams will often be active also during the dry season (Eriksen et al.).

A similar case of where climate and biodiversity measures have been coordinated is in the northern coast of Vietnam and described by Nguyen Hoang Tri and colleagues (1998). Preservation of the natural mangrove forest with its species diversity is important for local sources of income, for example, wood and honey production. The mangrove forest is equally important for protection from cyclones and typhoons. The coast is hit by between one and twelve typhoons per year, and there is great uncertainty about how the frequency and magnitude of these will be changed in connection with global warming. The mangrove forest protects agriculture against flood damage from cyclones and reduces the maintenance costs of the dikes. Preservation of this forest represents a type of climate adaptation. These examples illustrate that the interaction between various "global" environmental problems can result in different impacts in different places. To reduce the vulnerability of the poorest populations, policy measures should be adapted to local conditions and address several different environmental problems at the same time.

MEASURES UNDER THE CLIMATE CONVENTION

The Climate Convention requires developing countries to adapt public policy, government, and legislation to accommodate climate measures, both to reduce emissions of greenhouse gases and to adapt to possible climate changes. The Climate Convention from 1992 and the Kyoto Protocol from 1997 have two mechanisms that are particularly relevant with respect to implementing measures directed at developing countries. The Global Environment Facility (GEF) is the Climate Convention's funding scheme, designed to assist developing countries in developing climate strategies and integrating climate considerations into public policy, government, and development plans. The GEF is also supposed to fund preparation and implementation of practical measures, but until now no such GEF supported projects have been implemented. Most of the GEF's financial assistance has been directed at greenhouse gas emissions rather than enhancing local adaptation to climate impacts. As the poorest populations are responsible for very low greenhouse gas emissions but are nevertheless the most vulnerable to possible climate changes, it is thus a challenge to adapt the GEF to support measures designed to assist the poor. To truly integrate poverty considerations - which, for example, can be connected to local property rights - into environmental measures is also a great challenge because the GEF is essentially not a development fund but rather an environment fund. The GEF was established to fund measures specially directed at biodiversity and climate, and development measures are defined as outside its mandate.

The Kyoto Protocol's Clean Development Mechanism (CDM) is another important mechanism to finance measures in developing countries. This mechanism is designed to enable industrialized countries to meet their commitments to the Kyoto Protocol by financing development projects that reduce emissions of greenhouse gases in a developing country. Negotiations are still ongoing with respect to the regulatory framework of the CDM, but pilot projects have already been implemented. It is important that the CDM provide feasible solutions that benefit poor developing countries and that the CDM projects are not designed on the premises of the industrialized countries in a top-down fashion.

Another important source of funding for environmental measures in developing countries is development aid organizations such as NORAD or the World Bank. Since climate issues have previously only minimally been seen as an important development issue, only a few such organizations have paid attention to climate impacts. It will therefore be a challenge both to increase the awareness of the climate issue among these organizations and to enhance the cooperation between these organizations and the UN system in the design of policy measures. This is particularly important because development organizations have valuable experience with how locally adapted measures can be implemented to reach the poorest populations. The focus on poverty and local development is an important step forward in combating the negative impacts of climate change, but at the same time presents tremendous practical challenges. The question is how meaningful local solutions can be found to global problems.

GLOBAL ENVIRONMENTAL PROBLEMS

Trees and wild plants provide impoverished farmers in Kenya and Tanzania an alternative subsistence base in times of drought. These kind of local interactions tend to be ignored when climate changes are discussed at the global level.

The Intergovernmental Panel on Climate Change (IPCC) recently released a report that claims that the world's poorest populations will be the hardest hit by climate changes (IPCC 2001). The relationship between poverty and climate impacts has been known for some time - and poverty has often been connected to a lack of capacity to adapt to climatic changes (Adger & Kelly 2000). Moreover, poor people often live in risk-filled environments and subsist on natural resources that are negatively affected by climate changes. The IPCC's new emphasis on the poverty issue represents a fundamental shift in focus from climate impacts as a purely natural scientific phenomenon to climate impacts as a development issue. This shift has important consequences for the design and implementation of climate measures.

The planet is over-populated and its limited resources, unevenly distributed among the world's people, are being used at an accelerated pace. The implications of man's violence to the environment are serious throughout the world and particularly in less developed countries such as India. Air, water, land, and noise pollution constantly imperil quality of life and must be controlled and managed with all the skill at our disposal if man and his environment are to survive.

CLIMATE IMPACTS AND DEVELOPMENT ISSUES

Traditionally, climate has been seen as a global environmental issue, while poverty has been perceived as a local development issue. Thus there may be an initial contradiction between basing climate measures on international environmental agreements, such as the Climate Convention (UNFCCC), and taking poverty issues into account in these measures. Commitments, funding mechanisms, and policy measures established in the wake of climate agreements have been designed from the perspective of climate as a global environmental problem. Generally, a top-down approach to resolving climate issues, rather than a bottom-up approach has been taken. Some critics believe that perceiving environmental problems as global and designing global frameworks to address them implies a form of globalization that represents a new type of imperialism - one where the North gains control over natural resources and steers the development in the South under the guise of global environmental considerations (Mater 1999). Much of research into development issues represents the second basic view, namely that poverty is a development problem where the best solutions are shaped through local participation and policy measures are adapted to local conditions. In other words, this research takes a bottom-up approach to the problem, even though some of the basic causes of Siri Eriksen Cicerone 2/2001

GLOBAL ENVIRONMENTAL PROBLEMS AND LOCAL POVERTY

Poverty can be global (Chambers 1983). In terms of environmental research, there has been an increasing focus on the interrelationship between environment and development, and how nature and society affect each other through a complex interplay. This approach also implies that global environmental changes can have very different impacts on different local communities. Conservation of the baobab tree (Adansonia digitata) on cultivated land represents one climate adaptation in arid regions in East Africa. The fruit is a useful source of nutrition during drought years. The IPCC's suggestion that poverty plays an important role in determining how climate changes will affect various populations has two important consequences for policy design principles: The first is that those who are most vulnerable to the impacts of climate change should be the primary target of mitigation policy; poverty issues should be taken into account as a central element in climate measures. The second consequence is that policy design must take into account that the impacts of climate changes and possible effective measures to reduce vulnerability vary both from place to place and from time to time because social conditions change. It will be just as important to identify how adaptability can be improved among various social groups, as it will be to identify physical climate impacts. Measures should therefore be prepared in cooperation with the local population, where their priorities and conception of the problems become the center of focus.

Single problem, even though the causes of the degradation are considered to be separate phenomena at a global level. Environmental problems that are addressed separately at an international level - such as climate, biodiversity, and desertification - must at be tackled holistically at the local level. There are several practical examples of the interaction between several "different" environmental issues in locally adapted measures. One study in two arid agricultural areas (one in the Kitui district in Kenya and the other in the Same district in Tanzania) shows that biodiversity in the form of local tree and plant species distributed throughout the farms represents an alternative income source for poor farmers when crops fail. For example, local wood is used to make stools, kitchen equipment, chicken coops, and so on. These products are sold on the local markets. Lumber is also used to burn charcoal, which is sold to urban areas and cities. Leaves and seeds from certain trees are used as feed for goats and cattle. Drought resistant indigenous fruits are an important source of nutrition for both children and adults when there is little food. Thus preservation of local knowledge and biodiversity in cultivated areas helps enhance adaptability and reduce vulnerability to extreme climate events such as drought and flooding (Eriksen 2000). Preservation of the natural vegetation is also useful in combating desertification.

Moreover, particular ridges and hills covered with natural vegetation in dry areas have great value with respect to biodiversity. These areas often have different compositions of plant and animal species that otherwise are not found in the surrounding lowlands (Gachathi 1995). These ridges, which have higher precipitation than the lowlands, form important parts of local watersheds. In addition, there is a better microclimate (local temperatures and moisture) around these forested areas, particularly in arid regions that are dependent on the forests as water sources. Where ridge vegetation is well preserved, forest streams will often be active also during the dry season (Eriksen et al.).

A similar case of where climate and biodiversity measures have been coordinated is in the northern coast of Vietnam and described by Nguyen Hoang Tri and colleagues (1998). Preservation of the natural mangrove forest with its species diversity is important for local sources of income, for example, wood and honey production. The mangrove forest is equally important for protection from cyclones and typhoons. The coast is hit by between one and twelve typhoons per year, and there is great uncertainty about how the frequency and magnitude of these will be changed in connection with global warming. The mangrove forest protects agriculture against flood damage from cyclones and reduces the maintenance costs of the dikes. Preservation of this forest represents a type of climate adaptation. These examples illustrate that the interaction between various "global" environmental problems can result in different impacts in different places. To reduce the vulnerability of the poorest populations, policy measures should be adapted to local conditions and address several different environmental problems at the same time.

In developing countries seeking to expand their economic activities, consideration for environmental conservation often receives a low priority. In addition, approaches used in industria-lized countries often cannot be applied directly in developing countries. In this context, NIES is conducting research on ways to conserve the environment that are appropriate for developing country conditions.

Water quality and air pollution are serious problems in developing countries in the Asian region. Air pollution in major cities marked by many factories and heavy vehicle traffic also has high concentrations of sulfur dioxide and suspended particulate matter (SPM), at levels Japan experienced in the past. In addition, problems such as damage from acid rain and transboundary pollution are growing more serious. Pollution of rivers and lakes from chemical substances (including agricultural chemicals) and eutrophication (including abnormal growth of toxic algae) are also occurring more frequently, while water shortages and tropical forest destruction are worsening. While many developing countries give economic development the greatest priority, many problems remain with basic needs such as safe drinking water and food, as well as medical and public health services. This situation often hinders progress in addressing environmental problems. In some countries, including Bangladesh, China and India, negative health impacts are growing over large areas due to fluorine and arsenic pollution in air and drinking water implementing measures directed at developing countries. The Global Environment Facility (GEF) is the Climate Convention's funding scheme, designed to assist developing countries in developing climate strategies and integrating climate considerations into public policy, government, and development plans. The GEF is also supposed to fund preparation and implementation of practical measures, but until now no such GEF supported projects have been implemented. Most of the GEF's financial assistance has been directed at greenhouse gas emissions rather than enhancing local adaptation to climate impacts. As the poorest populations are responsible for very low greenhouse gas emissions but are nevertheless the most vulnerable to possible climate changes, it is thus a challenge to adapt the GEF to support measures designed to assist the poor. To truly integrate poverty considerations - which, for example, can be connected to local property rights - into environmental measures is also a great challenge because the GEF is essentially not a development fund but rather an environment fund. The GEF was established to fund measures specially directed at biodiversity and climate, and development measures are defined as outside its mandate.

The Kyoto Protocol's Clean Development Mechanism (CDM) is another important mechanism to finance measures in developing countries. This mechanism is designed to enable industrialized countries to meet their commitments to the Kyoto Protocol by financing development projects that reduce emissions of greenhouse gases in a developing country. Negotiations are still ongoing with respect to the regulatory framework of the CDM, but pilot projects have already been implemented. It is important that the CDM provide feasible solutions that benefit poor developing countries and that the CDM projects are not designed on the premises of the industrialized countries in a top-down fashion. Another important source of funding for environmental measures in developing countries is development aid organizations such as NORAD or the World Bank. Since climate issues have previously only minimally been seen as an important development issue, only a few such organizations have paid attention to climate impacts. It will therefore be a challenge both to increase the awareness of the climate issue among these organizations and to enhance the cooperation between these organizations and the UN system in the design of policy measures. This is particularly important because development organizations have valuable experience with how locally adapted measures can be implemented to reach the poorest populations. The focus on poverty and local development is an important step forward in combating the negative impacts of climate change, but at the same time presents tremendous practical challenges. The question is how meaningful local solutions can be found to global problems.

For a long time man has assumed that the environment enveloping his existence was in vulnerable, bountiful, and immense in its capacity to support life. Events of the last two centuries have belied these presuppositions. The exuberance of the new era of technology and abundance has evaporated and the bright prospects of economic progress and superabundant living have been shattered by the discovery of the finiteness of the planet. With all the technological developments, the world still suffers from the scourges of poverty, disease, and misery, along with environmental degradation, destruction of forests, and extinction of many species. The basic fact of this planet is that it is overpopulated. An increasing number of people are becoming urbanized, they continue to consume resources for survival, and to utilize them at an accelerated pace. By the end of the century, the population will swell to about 6.5 billion, two-thirds of whom will be living in developing countries. The prospect of supporting such a vast human population on the planet seems stupendous and grim, making the future of man unpredictable. It is now clearly recognized that urban man is continuously destroying many useful elements of life by using them indiscriminately and, consequently, degrading the environment which sustains life. It has been scientifically established that the ecosystem of the Spaceship Earth is fragile, limited, and that man's tampering with it may, in the end, make the earth unlivable, not only for man but for all life forms. Man has begun to live in large agglomerations of millions of people; he has command of technologies, which can take him to the farthest planets in the solar system; and he has unraveled mysteries that were secrets of nature only 100 years ago. In a sense, the twentieth century could have been a turning point in man's consciousness about himself and his universe, which should have freed the human race from the perennial threats of destitution, privation, and death. In reality, however, urban life in many parts of the world has become an insult to mankind because of the total degradation of people and their environment. In this context, UNESCO's "Man and the Biosphere Program" (MAB) states its objective to be "development of a basis through natural and social sciences for the rational use and conservation of the resources of the biosphere and improvement of the relation between man and the environment." Further, it states that "the consequences of today's actions on tomorrow's world should be predicted and man must manage natural resources efficiently." This report succinctly underscores the uncertain future of man if the symbiotic relationship between man and his environment continues to deteriorate at the present rate.

All through the two million years of man's existence on the planet, his survival has been conditioned by the intricate balance among the various elements of the earth, which have provided a continuous supply of materials for his needs. During the last 300 years, how ever, man's relationship with the earth, its environment, and outer space has changed drastically be cause of the accumulation of vast knowledge, increasing technology, exponential increase in expectations and needs, and, finally, a quantum jump in human activities by exploding population.

The techno logical revolution has ushered in a new era of transformation, which has not left any part of, the biosphere untouched, directly or indirectly, making man's survival increasingly precarious. Urbanization, as well, has resulted in a new order of relationships between human society and the ecosystem.

The Earth went through a series of drastic changes for billions of years until, ultimately, life originated, leading eventually to the evolution of man. Only recently, the complex web of human activities has generated extensive modifications of the environment. Every moment man is displacing millions of tons of various elements from one area to another, trans-forming them from one form to another, and converting them into a. large number of products through different processes. For example, 800 million tons of various metals are taken from the earth, 40 million tons of toxic chemicals, and 7 billion tons of conditional fuel units are consumed every year.

In short, the immense number of industrial activities of man is producing vast quantities of pollutants, which result in massive degradation of the environment. An alarming threat to the life-preserving ecosystem is evident around the globe. It can be observed in acid rains, disappearance of plant and animal species, desertification, smog, destruction of coastal areas, dirty waterways, poisoned foods, and blaring noises. Looking at the plight of the enduring earth, what Barry Commoner once said seems to be true: "We have been living under a vast and potentially fatal illusion: that we can enjoy the enormous benefits of modern technology with out risk to the integrity of human life and the environment." Slowly, land, air, and water are being vandalized by man's vast, indiscriminate plundering of the environment. How far the ecosystem of the earth will be able to bear the threat before collapsing is only a question of time. Exploding population, increasingly complex technology, and careless use of resources have joined to face us with a triple danger. Unless man finds another suitable environment or creates a new one, or unless he adapts to the increasingly polluted environment, he must put the planet in order by restoring natural equilibrium if he does not want to face extinction.

The main reason is that in our society the environment has become a scarce resource. Since economics is about how to tackle scarce resources, it can often be useful when dealing with environmental problems. One way of using economics is to ensure that the costs and the benefits of environmental measures are well balanced. Although it is difficult to estimate costs and benefits, there is an increasing demand that this is done before new environmental policy is decided. With the use of market-based instruments, environmental goals can sometimes be reached more efficiently than with traditional command and control regulations.

Through the introduction of new regulations or the amendment of existing regulations, environmental and other policies may result in impacts across society. These can include changes in government expenditure, impacts on different sectors of industry, changes in costs faced by consumers and a range of other social and environmental effects. Within the policy making process, there is a range of approaches which could be adopted to provide decision makers with an understanding of the implications of a policy/regulation. These range from qualitative approaches which rely on a description of the impacts to quantitative approaches which attempt to provide further detail on the nature, magnitude and significance of potential effects. Economics provides two approaches which fall into the latter category, these being cost-benefit analysis and cost effectiveness analysis:

1. Cost-benefit analysis examines the trade-offs in terms of the costs and benefits of a policy; and
2. Cost-effectiveness analysis determines the least-cost option of attaining a pre-defined target.

In some countries, these two techniques are applied across many policy areas, while in others they are used more commonly in particular areas, such as environmental regulation, than in others. However, application in general varies greatly across EU countries and worldwide. In many cases, this variation is a reflection of different cultural perspectives, views on risk and uncertainty, and concerns over the need to ensure a more sustainable future.

During the early part of this century, and particularly the period following World War II, an increasingly regulatory approach to government developed in many western democracies. This was in part a response to rapid technological developments, leading to an expectation that governments should intercede in the resulting changes to society and business. By the time the oil crisis of the early 1970s hit the international economy, there was significant support for the belief that regulation had become excessive. In the midst of a global recession, the need to spend limited government resources to achieve the greatest social good coincided with a push in many western countries to curtail regulations and thus unfetter businesses from needless regula-tory constraints. In the push for deregulation, questions developed about which regulations to remove, which to maintain, and on what basis should such decisions be made. The criteria for judging whether a regulation is necessary or excessive ranged - and still range - from economic to political, ethical and cultural and other social, scientific and technical criteria. The focus of this book is on the use of economic criteria for policy and regulatory analysis. It must nevertheless be acknowledged that the other, non-economic criteria influence both the content of economic appraisals and the way in which they are received.

RISK & POLICY ANALYSTS

Given that this book is primarily concerned with economic appraisal of environmental policies and regulations, it is also important to note that the past 30 years have witnessed a growing awareness of the impacts of human activities on the environment. Whether the issue is diminishing fish stocks in rivers, smog or acid rain, economists have been exploring how to improve the environment while accommodating economic growth. The resulting efforts have contributed to the evolution of the push for deregulation into one for better regulation. It is notable that the use of economic appraisal tools appears to have developed most in those countries, which most embrace free market principles, with science-based tools being preferred in other countries.

APPRAISAL TOOLS

Although the focus of this book concerns the degree and manner to which different countries apply CBA and CEA, it is important to remember that there is a range of different assessment tools which could be used to assist in the appraisal of policies and government project-related expenditure. The use of CBA and CEA must, therefore, be seen within the context of appraisal more generally, where other potential tools include:

1. Compliance cost analysis;
2. Environmental Impact Assessment (EIA);
3. Simple scoring and weighting techniques and more sophisticated Multi-Criteria Analysis (MCA);
4. Trend analysis; and
5. Checklists.

All of these tools provide some indication of the advantages and drawbacks of a given proposal. The key variations are in the level of detail required, the degree of quantification involved in applying the methodology and the area of focus.

2

Sustainable Development

The concept of sustainable development is now considered a guiding principle of national and international action. Yet the widespread acceptance of this concept stands in contrast with the inability so far to alter effectively the development model responsible for environmental degradation. The lack of many positive and concrete results produced by massive efforts in the field of international cooperation for the environment indicate the contradictory character of this new "global" environmentalism.

The purpose of this chapter is to explore how environmental considerations were reframed so as to become compatible with global development. Adopting an international political economy perspective and based on interviews with the main categories of actors involved, it provides evidence that environmental concerns were remodelled by the joint action of technocratic environmentalists, the international UN-related development establishment and business and industry sectors. Analyzing the results of international cooperation and in particular the review of UNCED's implementation five years after the Summit, the article questions the nature of the "sustainable development" consensus. The inability of the international community to deal with most global environmental issues reveals the limits of international cooperation in the name of the environment.

A significant feature of international politics since the end of the 1980s has been the growing concern with environmental protection and the multiplication of the number of international conferences and agreements in this area. Environmental protection is presently recognized as a major political issue, and has acquired a well-defined position on the international political agenda. The United Nations Conference on Environment and Development (UNCED) held in Rio de Janeiro from 3 to 14 June 1992, was a unique moment in diplomatic history. The conference heralded the most elaborate attempt ever to develop institutional solutions to major environmental problems.

Based on the idea that "environment" and "development" had to be linked in a comprehensive framework that would allow for the generalization of economic growth and prosperity while including environmental concerns, UNCED came out with a global solution to the ecological crisis, the concept of "sustainable development." A global bargain was struck, according to which developed nations would provide some financial resources and transfer appropriate and "clean" technology to developing countries to help them protect their environments. An international mechanism - the Global Environment Facility (GEF) - was settled on to undertake the funding of international projects.

At the same time, global conventions on Climate Change and Biological Diversity were negotiated in an attempt to control the most devastating effects of economic activities, such as CO emissions from industry and consumers, and to protect the earth's living capacity. A program of action, "Agenda 21," was carefully worked out, covering all areas from health to institutions, from the role of women to the responsibilities of business, all in order to serve as a guide for action to attain sustainability in every country. To facilitate the transition towards "sustainable development," developed countries promised large sums of money in the form of aid, investment and pollution control projects.

The Conference generated a high degree of optimism as to the international community's ability to deal with global environmental problems. Development could continue, now on a truly global base, without the risk of the complete exhaustion of natural resources or of other major environmental catastrophes. The Cold War was over, and rational planning, technology and economic instruments would ensure the extension of the capitalist model of accumulation worldwide.

BASIC PRINCIPLES OF SUSTAINABLE DEVELOPMENT

Great ideas are usually simple ideas. While the specific analysis of any important topic will necessarily involve complexity and subtlety, the fundamental concepts which underlie powerful paradigms of thought are usually relatively straightforward and easy to grasp. In the area of social science, ideas which affect millions of people and guide the policies of nations must be accessible to all, not just to an elite. Only thus can they permeate institutions from the local to the global level, and become a part of the human landscape, part of the fabric within which we define our lives. Such is the concept of development. Prior to the second half of the twentieth century, the idea of development as we know it today barely existed. The structures of imperial and colonial power which dominated the world in the nineteenth and early twentieth centuries made little provision for economic and social advance in what we now call the developing world. Colonial regions functioned primarily to supply imperial powers with raw materials and cheap labour – including slave labour as late as the mid-nineteenth century.

Within the richer countries of Europe, North America, and Japan, economic growth was of course central to the generally accepted goals of "progress" and "modernization", but there was relatively little concern for issues of equity and social justice. The desperate poverty and weak or non-existent social safety nets in Europe and the United States during the Great Depression showed how even in these countries, policy was not driven by the needs of the majority of people.

By the end of the Second World War, perceptions and policy had changed drastically. Economic and social improvement for the majority had become a major preoccupation of governments, and with the crumbling of colonial power relations this goal was extended to the poorer nations of the world. Economic development, with its social and institutional correlates, came to occupy an essential place in theory and policy, as well as in the Cold War competition between capitalism and communism. As the historian of economic thought Roger Backhouse puts it: Development economics in its modern form did not exist before the 1940's. The concern of development economics, as the term is now understood, is with countries or regions which are seen to be *under* or *less* developed relative to others, and which, it is commonly believed, *should*, if they are not to become ever poorer relative to the developed countries, be developed in some way.

Within formal neoclassical economic theory, an effort has been made to achieve a positive rather than a normative perspective – that is, to describe what *is* rather than positing what *should be.* Development economics, in contrast, is explicitly normative, as Backhouse's description makes clear. As such, it cannot avoid concern with social and political issues, and must focus on goals, ideals, and ends, as well as economic means.

When W.W. Rostow published his ambitious overview of economic development, The Stages of Economic Growth, in 1960, he subtitled it "A Non-Communist Manifesto". Conscious of the claims of Marxism to offer a path to a better future for the majority of the world's peoples, Rostow sought to counterpose a superior vision of social and economic goals. Notable in this perspective was a linear conception of economic development.

According to this view, all successfully developing countries would pass through a series of stages, from traditional society through economic "take-off" to maturity and high massconsumption. The "less-developed" nations therefore might reasonably hope to achieve the "mature" status of the U.S. and Europe without the need for communist revolution. Rostow's concept of take-off, as well as his overall perspective of economic and social progress towards a goal of mass consumption, was widely accepted by development theorists.

Thus economists, other social scientists, and policymakers adopted a framework of thought which was much more ambitious in its scope than previous formulations of political economy. The clear goal of economic development policy was to raise living standards throughout the world, providing steadily more goods and services to an expanding population. The international institutional structures set up after the second world war, including the International Monetary Fund, the World Bank, and the United Nations, were specifically designed with this goal in mind.

As development policy has evolved, different approaches have been emphasized at different times. The original emphasis was on promoting more productive agriculture and industrialization. In the late 1970's a focus on basic needs was advocated by Paul Streeten, Mahbub Ul Haq, and others. Education, nutrition, health, sanitation, and employment for the poor were the central components of this approach – reflecting an acknowledgment that the benefits of development did not necessarily "trickle down" to those who needed them most. This perspective inspired the creation of the United Nations Development Programme's Human Development Index, which uses health and education measures together with Gross Domestic Product (GDP) to calculate an overall index of development success.

In the 1980's the focus shifted to "structural adjustment", including liberalization of trade, eliminating government deficits and overvalued exchange rates, and dismantling inefficient parastatal organizations. Structural adjustment was seen as correcting the errors of earlier, government-centred development policies which had led to bloated bureaucracies, unbalanced budgets, and excessive debt. But critiques of structural adjustment policies have found them at odds with the basic needs priorities. Market-oriented reforms have often lead to greater inequality and hardship for the poor even as economic efficiency improved. A tension thus remains between the basic needs and market-oriented perspectives on development.

At the turn of the century, what is the 50-year record of the broad-reaching, and historically fairly young, effort at global development? The concept has been widely accepted, by countries of varied political structure. There have been remarkable successes – notably in East Asia – and worldwide progress both in standard GDP measures and in measures of human development such as life expectancy and education.

There have also been areas of slow or negative growth, especially in Africa, where GDP increase was slow and food production per capita in decline even before the rapid spread of AIDS devastated many countries and dramatically lowered life expectancies. Globally, most countries have made significant advances both in GDP and in Human Development Index measures. But overall, the record of development on a world scale is open to two major criticisms:

- The benefits of development have been distributed unevenly, with income inequalities remaining persistent and sometimes increasing over time. The global numbers of extremely poor and malnourished people have remained high, and in some areas have increased, even as a global middle class has achieved relative affluence.
- There have been major negative impacts of development on the environment and on existing social structures. Many traditional societies have been devastated by development of forests, water systems, and intensive fisheries. Urban areas in developing countries commonly suffer from extreme pollution and inadequate transportation, water, and sewer infrastructure. Environmental damage, if unchecked, may undermine the achievements of development and even lead to collapse of essential ecosystems.

These problems are not minor blemishes on an overall record of success. Rather, they appear to be endemic to development as it has taken place over the past half-century, and to threaten to turn success into failure. World Bank President James Wolfensohn and chief economist Joseph Stiglitz acknowledged in 1999 that these issues are crucial to address if global development is to succeed. Harsher critics of the development paradigm, such as Richard Norgaard, see them as indicative of fundamental error:

Modernism, and its more recent manifestation as development, have betrayed progress... while a few have attained material abundance, resource depletion and environmental degradation now endanger many and threaten the hopes of all to come... Modernism betrayed progress by leading us into, preventing us from seeing, and keeping us from addressing interwoven environmental, organizational, and cultural problems.

Whether we seek a reform or a radical rethinking of the concept of development, it is evident that changes are required in both goals and methods.

The straightforward view of development as an upward climb, common to all nations but with different countries at different stages, seems inadequate for the twenty-first century. The absolute gaps between rich and poor nations, and between rich and poor groups within individual countries, are widening, not narrowing. And even if we can imagine all nations reaching stable populations and satisfactory levels of GDP by, say, 2050, can we envision the planetary ecosystem surviving the greatly increased demands on its resources and environmental absorption capacity?

The growing awareness of these challenges to traditional development thinking has led to the increasingly wide acceptance of a new concept – that of sustainable development. Development which protects the environment, development which advances social justice phrases such as these have surrounded the introduction of what has been claimed to be a new paradigm. The new formulation has been eagerly adopted both by critics of standard development practice and by leaders of existing development institutions. But what does sustainable development really mean?

Defining a New Paradigm

When the World Commission on Environment and Development presented their 1987 report, Our Common Future, they sought to address the problem of conflicts between environment and development goals by formulating a definition of sustainable development: Sustainable development is development which meets the needs of the present without compromising the ability of future generations to meet their own needs.

In the extensive discussion and use of the concept since then, there has generally been a recognition of three aspects of sustainable development:

- *Economic:* An economically sustainable system must be able to produce goods and services on a continuing basis, to maintain manageable levels of government and external debt, and to avoid extreme sectoral imbalances which damage agricultural or industrial production.
- *Environmental:* An environmentally sustainable system must maintain a stable resource base, avoiding over-exploitation of renewable resource systems or environmental sink functions, and depleting non-renewable resources only to the extent that investment is made in adequate substitutes. This includes maintenance of biodiversity, atmospheric stability, and other ecosystem functions not ordinarily classed as economic resources.
- *Social:* A socially sustainable system must achieve distributional equity, adequate provision of social services including health and education, gender equity, and political accountability and participation.

Clearly, these three elements of sustainability introduce many potential complications to the original simple definition. The goals expressed or implied are multidimensional, raising the issue of how to balance objectives and how to judge success or failure. For example, what if provision of adequate food and water supplies appears to require changes in land use which will decrease biodiversity? What if non-polluting energy sources are more expensive, thus increasing the burden on the poor, for whom they represent a larger proportion of daily expenditure? Which goal will take precedence?

In the real world, we can rarely avoid trade-offs, and as Richard Norgaard points out, we can "maximize" only one objective at a time. Norgaard concludes that "it is impossible to define sustainable development in an operational manner in the detail and with the level of control presumed in the logic of modernity." The strongly normative nature of the sustainable development concept makes it difficult to pin down analytically.

Nonetheless, the three principles outlined above do have resonance at a commonsense level. They satisfy the criterion set forth earlier for a powerful, easily grasped concept which can have wide applicability. Surely if we could move closer to achieving this tripartite goal, the world would be a better place – and equally surely we frequently fall short in all three respects. It may be easier to identify un-sustainability than sustainability – and the identification of un-sustainability can motivate us to take necessary policy action.

It is instructive to examine the problem from different disciplinary perspectives. Certainly the goals set forth require the insights of multiple disciplines. Economists, one might assume, would tend to give greater weight to the economic objectives, ecologists to the environmental dimension, and social theorists to the social issues. But before we can attempt to balance these different perspectives, we need to understand them and explore their internal logics.

Each of the three areas is commonly referred to as a *system*: economic systems, environmental systems, and social systems each have their own logic. It is an impossible task to analyse all these systems at once. Therefore we must start by considering each separately, as suggested by the Balaton Group's report on sustainability indicators: The total system of which human society is a part, and on which it depends for support, is made up of a large number of component systems. The whole cannot function properly and is not viable and sustainable if individual component systems cannot function properly...sustainable development is possible only if component systems as well as the total system are viable. Despite the uncertainty of the direction of sustainable development, it is necessary to identify the essential component systems and to define indicators that can provide essential and reliable information about the viability of each and of the total system.

This implies that we can use different indicators to measure different dimensions of sustainability. Indicators imply measurement; measurement implies the theoretical definition of concepts to measure. Let us examine what the three different disciplinary areas have to offer in this regard.

PRESENT CHALLENGES TO SUSTAINABLE DEVELOPMENT

There is no doubt though that significant progress has been made since the first "Brandt Report": We saw positive political developments with the end of the cold war, the decline of the nuclear threat and a significant decrease in global military spending; a longer life expectancy, lower infant mortality rates, and higher educational achievements indicate some progress in social development; the "East Asian Miracle" told us economic success stories. Last but not least, the "governance" issue is no longer treated on the quiet in the international development debate. Hence, many battles for development have been fought and won-but by far not the war: Comparing the range of development deficits listed in the first "Brandt-Report" with those listed in the 1994 Human Development Report in the World watch Institute Report on Progress Toward a Sustainable Society, we still find a persistent increase in real numbers of those living in poverty, ever stronger population pressures and environmental degradation as well as largely unmet human security needs in political, economic, nutritional, health, environmental, human rights and personal safety terms. As long as there are more than 1,300 million people living in absolute poverty, the same number of people lacking access to clean water and nearly two billions to safe sanitation, some 800 million hungry people, millions of premature deaths due to poor nutrition and polluted water, thousands of hectares of arable land and forests being destroyed every day and an unknown number of species being eradicated every day, Sustainable human development remains just a utopian dream. Given the interdependencies of the different factors that combine against a sustainable human development, it is arbitrary to single out one specific problem area and deal with it in isolation.

THE ECOLOGICAL CONSEQUENCES OF RAPID POPULATION GROWTH

Over the last 15 years, around 100 million people have been added annually to the world's population. Today we are more than 5.6 billion people on earth. Well over 90% of this population growth occurs in developing countries, where it poses serious problems to endeavors to raise people's living standards and to provide them with basic health and educational services as well as other basic needs items. Past and present rates of population growth have also had a variety of environmental consequences. The most important amongst these are the destruction of fertile soils and forests. International evidence suggests, that it is possible to lower population growth in an ethically acceptable way. The main pillars of a development policy that motivates parents to have smaller families are the satisfaction of basic needs - including the availability of and access to appropriate family planning services, economic, political and social reforms to create distributional equity and equality of opportunities, especially for women; and good governance resulting in social security and guaranteed human rights. The traditional explanation for human suffering in the developing countries is that these countries are simply too poor to pay for a minimum of infrastructure and services to meet the basic needs of their poorer classes.

In June 1991 the United Nations Development Program (UNDP), in its report on Human Development for 1991, presented an equally relevant conclusion: "The lack of political commitment, not of financial resources, is often the real cause for human neglect." The Report concludes that much current spending is misdirected and inefficiently used. If the priorities are set right, more money will be available for accelerated human progress. Here we have an official United Nations statement which says bluntly what development experts have been complaining about for years: the failure of politicians in many if not most developing countries is a fundamental obstacle to sustainable development. There are too many wrong priorities, too much wastage of resources, too high (and sometimes rising) expenditures for military purposes, too many inefficient public enterprises, too many prestige projects, too much capital flight and proliferating corruption.

Commonwealth General Secretary Shridath Ramphal once called the "trilogy of imperfection", as it namely,

1. Rampant concentration of political power and/or dismantling of democratic structures,
2. Excessive military expenditures, and
3. Chronic corruption.

Weakens any sensible development and population strategy in its earliest stages and makes lasting economic and social improvements impossible. If priorities were set properly and corresponding budgetary changes made, it would be possible within a short period of time - and without a single extra dollar of development assistance-to make available fifty billion US dollars for the fight against human misery - for example, for elementary education, basic health services, rural water supplies, family planning services, food programs and social security. Even politically conservative institutions like the World Bank have now acknowledged the substantial and lasting success of development strategies oriented toward basic needs - so we are not engaging in speculation. Of course many developing countries will continue in the future to need financial and technical support from the rich countries of the world. Many developing countries, and hence a large part of the world's population, are too deeply mired in poverty and - even after the requisite house-cleaning in terms of development policy - have too few resources to enter onto an appropriate path of human development. In the future, however, the best argument for making more development assistance available to a country will be the rational expenditure pattern of its own already available resources. Suitable national budget priorities and appropriate international development cooperation will produce different project concepts and programs from one country to another, but in the poorest countries strategies oriented toward basic needs, including family planning programs, will be in the forefront, along with support for rural infrastructure, strengthening the role of women, and ecologically acceptable programs for modernizing agriculture.

Therefore, population policy measures which in varying degrees put pressure on parents can only be justified if the socio-political preconditions for voluntary methods have been properly and adequately tried and have nevertheless failed or are bound to fail in the future, something which, until now at least, has not been proved. Besides a population policy, there are a number of further ecological policy components for developing countries, be it the promotion of a sustainable rural and agricultural development, managing fragile ecosystems or other programmes dealt with in Agenda 21. Many of them, however, must be initiated and implemented first and foremost by the industrial countries. They are not only the main polluters of this globe but - through global communication networks and the consumption culture spots carried by them - also shape the model through which "progress" is defined.

Destruction of Fertile Soils

One of the most serious consequences of destroying tropical forests, particularly rain forests, is soil erosion. Erosion means that the productive topsoil, deprived of vegetation, is carried away by wind or water. In eroded areas, the process of desertification is able to run its course. There are a number of conditions related to climate, topography, and composition of the soil and nature of vegetation, which trigger natural erosion. The human action, which intensifies erosion, is generally the destruction of stabilizing vegetation. Wind erosion produces its worst effects in arid and semi-arid areas, for example those that have been deprived of their natural cover by overgrazing. More than 22 percent of the land area in Africa north of the equator and 35 percent in the Near East is exposed to wind erosion. Under extreme conditions up to 150 tons of soil can be carried away from a hectare of land in one hour. Soils on sloping land and in mountainous regions, which lose their stabilizing cover of vegetation through improper agricultural methods, are exposed to water erosion. Soil is washed away into rivers by precipitation (or glaciers) and accumulates there as silt which, by raising the water level, causes flooding. Deforestation in the Himalayas, for example, causes a considerably larger run-off than the lower-lying areas in the valleys can accept without damage. The area of India hit by severe flooding has more than tripled since 1960. Flooding was so bad in the fall of 1988 that two-thirds of Bangladesh was under water for several days and the rice crop suffered such big losses that substantial amounts of grain had to be imported.

Because of strong population pressures and inequitable land distribution, the humid tropics have the highest rate of soil erosion and the heaviest burden of river sedimentation in the world. In this connection, the Enquete Commission of the German Bundestag gives special mention to investigations of the Ambuklao and Binga Rivers in the Philippines and the Brahmaputra in India. The Yangtze in China and the Ganges in India are said to carry off about 3 billion tons of soil a year. Between seven and nine million hectares of arable land - the estimates vary - are destroyed every year, as a result of erosion, flooding, Stalinization, or overuse. A further 200,000 square kilometers lose practically their entire productivity. The Food and Agriculture Organization of the United Nations (FAO) fears that a total of approximately 544 million hectares - or 18% of the arable land in the world - could be lost forever if no measures are taken to preserve them. In areas where soils are not being directly destroyed, the fertility will decline if current patterns of use are continued. As a result of erosion alone, i.e. the removal of the top layer of soil and its organic components through water and wind, 29% of the food production in rain fed fields are threatened.

Soil erosion caused by water is one of the most serious problems in the tropics and takes on a magnitude in many places that can no longer be offset by soil regeneration measures. India loses 16.35 tons of soil substance per hectare per year in this manner, and only 4.5 to 11.2 tons can be regenerated using appropriate cultivation measures. Sixty one percent of the eroded materials are deposited elsewhere, 10% remain in reservoirs (reducing dam capacities by 1-2% per year), and 29% (1.5 billion tons per year) are lost forever to the ocean. The U.N. Population Fund points to the fact that "unchecked soil erosion could well cause a decline of 19 to 29 per cent in food production from rain fed croplands during the 25 years from 1985 to 2010." Erosion can be observed particularly in areas where marginal land is being farmed (e.g. on slopes), where soil is exposed without protection to wind and water (e.g. as a result of cleared forests or the removal of bushes and hedges), and where soils are used beyond their regeneration capacity. The associated losses in productivity are substantial; the reduced capacity of the soil to absorb water and thereby to prevent downstream flooding is enormous. The overuse of agricultural production areas is a consequence of various factors, one of which is poverty. Impoverished peasants cannot afford the conservation measures needed to protect the soil.

Another factor is growing population pressure. Where the number of people rises, there is either more pressure on the land or additional land is taken into agricultural production. In the first case, farming becomes more intensive, fallow periods are shortened. The results are exhaustion and degradation of land. Where an increased number of people must expand acreage to areas that are suitable only for extensive cattle raising, forcing the cattle farmers, in turn, to move to areas that are unsuitable for cattle raising, the situation is particularly serious. This has not only negative ecological consequences but also undermines food security and subsistence for a large part of the population. Various population and environment experts such as the U.N. Population Fund use the example of Java to portray the consequences of the overuse of agricultural land:

> *"The population has surged from 51 million in 1950 to 112 million today; 62 per cent of the nation's total population is now located on 8 per cent of its national territory. This rapid buildup in human numbers has served to aggravate soil erosion. In 44,000 square kilometers of upland farming areas, the population density has reached a level of 700 to 900 persons per square kilometer, even 2,000 persons or more in some localities, while the average land holding has declined to a mere 0.7 of a hectare. On third of these upland areas are seriously eroding, and more than 10,000 square kilometers of grain lands have been degraded to the point they no longer support even subsistence farming. This threatens the livelihood of 12 million people, many of whom live in absolute poverty and have no means to engage in soil-conservation practices."*

The area of Nepal that borders on India is confronted by similar problems; there, too, continuing population pressure has forced the people to use steeper areas for agricultural purposes. In some cases the cultivation of mountain slopes becomes possible only after forests are cleared, which, in turn, causes further erosion. High population growth (2.5% in 1992) and existing inheritance laws have caused the average area of cultivation for one household to drop from 3-4 hectares in the sixties to approximate one hectare today. Already today one third of all households is no longer in a position to live off of its own land.

Desertification or the transformation of agricultural land in semi-arid and arid regions into non-cultivable land due to overuse by humans - is even a more severe form of land degradation. UNFPA estimates this process to be threatening:

> *" 45 million square kilometers or a full third of the Earth's land surface - together with the livelihoods of at least 850 million people, of whom 135 million are experiencing the rigors of severe desertification. Already it eliminates 60,000 square kilometers of agricultural land each year, and impoverishes another 200,000 square kilometers, reducing yields and requiring costly remedial measures. The costs in terms of agricultural output foregone are estimated to be in the order of USD 30 billion a year. One of the main causes of desertification is over-grazing by domestic livestock."*

Today, desertification processes have advanced everywhere where traditional, soil-preserving farming methods were abandoned due to increasing population pressures, and new methods were introduced which cannot be applied over long periods. Inappropriate agricultural policies, disregard for the needs of small farmers, as well as an absence of resource management and other factors have helped to bring about these problems, and this in regions that already were at a considerable disadvantage in terms of climate and natural conditions. Therefore, population policy measures alone cannot prevent the destruction of fertile soils. A slowing of population growth, however, could give the respective countries a breather and could substantially alleviate development policy efforts of all kinds.

THE DESTRUCTION OF FORESTS

The frightening dimensions of contemporary forest destruction are indisputable. In the last 45 years destruction of soil and forests has damaged approximately 11 percent of the earth's vegetation so seriously that the biological functions have been lost. Today, less than half of the original stands of tropical forest remain. The destruction of tropical rain forests causes special concern. According to a report of the World Resources Institute in Washington, it progressed between 1981 and 1990 at an average rate of about 17 million hectares (ca. 42 million acres) a year, considerably faster than has hitherto been assumed. Wherever population growth is particularly high (above 2.5 percent per year), wherever a large part of the population lives in absolute poverty, wherever unjust arrangements for land ownership and leasehold are to be found or other structures exist which hinder development, the destruction of forests takes on especially dramatic proportions. Here again, therefore, it is not just population growth, which creates the problems but its combination with other problems of underdevelopment.

The causes of the destruction of tropical forests - as those of the destruction of fertile soils - are quite complex. The kinds and causes of damage vary in significance from one country to another and the speed with which the destruction progresses depends on the social, economic, or political conditions that prevail in the respective countries (e.g. support of small farmers, the importance of agriculture for employment and national income, issues of land reform). Among the most important causes are:

- Shifting cultivation mainly or exclusively for subsistence;
- Agro-industrial land use for cash crop production (e.g. coffee or palm oil) or fodder (e.g. soybeans or corn); extensive animal husbandry, chiefly cattle;
- Cutting firewood; lumbering;
- Clearing land for mining or power supply (dams).

Along with the damage to soils through overuse, the destruction of tropical forest areas is largely a direct result of growing population pressure and of the consequent need to expand agricultural land, estimates of the United Nations' Population Fund set the figure at 79 percent.

- In the Philippines, for example, the combination of population pressure and scarce land led to the defores-tation and settlement of the uplands with slopes as steep as 45 degrees. Extreme poverty, insecurity of tenure in their farmlands, lack of agricultural extension services and credit institutions prevented the settlers from investing in means and methods of soil conservation. The result has been devastating erosion and a growing water shortage.
- Ethiopia, one of the poorest countries in the world, faces the same problems. There, once fertile plateau land is losing between 1.5 and 3.5 billion tons of humus per year.

- In Kenya, population pressure on arable land and on forests has caused so much erosion that the "potential food output could eventually decline by as much as 50 per cent if soil loss cannot be reversed." In other regions of the developing world different causes predominate.
- In Costa Rica, for example, it is meat production, for which forest areas have been turned into pastureland.
- In West Africa and Southeast Asia projects of the wood industry are eating up the forest reserves.
- In Brazil, high unemployment and a widespread scarcity of agricultural land have provided the stimulus for clearing activities.

For a growing number of people, deforestation offers the only - also the quickest - way of securing a livelihood. The fact that the areas which small farmers clear for their subsistence can only be used for agricultural purposes for a very short time hastens the process of forest destruction. The fascinating variety of vegetation in tropical rain forests obscures the fact that their soils contain relatively little nourishment and that, as a result, there is a steep drop in yields after as little as two or three years. After that, new forest areas are cleared. If the cycles of rotation were long enough, i.e. if the land were permitted to lie fallow for an appropriate period of time (35-100 years) as was done in traditional patterns of shifting cultivation, then most of the negative consequences of clearance could be avoided. But under the pressure of growing populations this has become impossible. In many cases it is impossible to distinguish between clearing for lumber and clearing to obtain tillable agricultural land. Building roads through rain forests not only facilitates access for the felling of certain kinds of valuable wood but also eases the way for landless settlers to press forward into the rain forest and put large areas of virgin rain forest to agricultural use. The large-scale destruction of forests, especially tropical forests, has a number of serious ecological, social and economic consequences.

DECOUPLING ENVIRONMENTAL DEGRADATION AND ECONOMIC GROWTH

In the second half of the 20th century world population doubled, food production tripled, energy use quadrupled, and overall economic activity quintupled. Historically there has been a close correlation between economic growth and environmental degradation: as communities grow, so the environment declines. This trend is clearly demonstrated on graphs of human population numbers, economic growth, and environmental indicators.

Unsustainable economic growth has been starkly compared to the malignant growth of a cancer because it eats away at the Earth's ecosystem services which are its life-support system. There is concern that, unless resource use is checked, modern global civilization will follow the path of ancient civilizations that collapsed through overexploitation of their resource base. While conventional economics is concerned largely with economic growth and the efficient allocation of resources, ecological economics has the explicit goal of sustainable scale (rather than continual growth), fair distribution and efficient allocation, in that order.

The World Business Council for Sustainable Development states that "business cannot succeed in societies that fail". Sustainability studies analyse ways to reduce (decouple) the amount of resource (e.g. water, energy, or materials) needed for the production, consumption and disposal of a unit of good or service whether this be achieved from improved economic management, product design, new technology etc. Ecological economics includes the study of societal metabolism, the throughput of resources that enter and exit the economic system in relation to environmental quality.

NATURE AS AN ECONOMIC EXTERNALITY

The economic importance of nature is indicated by the use of the expression ecosystem services to highlight the market relevance of an increasingly scarce natural world that can no longer be regarded as both unlimited and free. In general as a commodity or service becomes more scarce the price increases and this acts as a restraint that encourages frugality, technical innovation and alternative products. However, this only applies when the product or service falls within the market system. As ecosystem services are generally treated as economic externalities they are unpriced and therefore overused and degraded, a situation sometimes referred to as the Tragedy of the Commons. One approach to this dilemma has been the attempt to "internalise" these "externalities" by using market strategies like ecotaxes and incentives, tradeable permits for carbon, water and nitrogen use etc., and the encouragement of payment for ecosystem services. Community currencies such as LETS, a gift economy and Time Banking have also been promoted as a way of supporting local economies and the environment. Green economics is another market-based attempt to address issues of equity and the environment. The global recession and a range of government policies that have been connected to that, are likely to bring the biggest annual fall in the world's carbon dioxide emissions in 40 years.

ECOSYSTEM SERVICES

Humankind benefits from a multitude of resources and processes that are supplied by natural ecosystems. Collectively, these benefits are known as ecosystem services and include products like clean drinking water and processes such as the decomposition of wastes. While scientists and environmentalists have discussed ecosystem services for decades, these services were popularized and their definitions formalized by the United Nations 2004 Millennium Ecosystem Assessment (MA), a four-year study involving more than 1,300 scientists worldwide. This grouped ecosystem services into four broad categories: *provisioning,* such as the production of food and water; *regulating,* such as the control of climate and disease; *supporting,* such as nutrient cycles and crop pollination; and *cultural,* such as spiritual and recreational benefits.

As human populations grow, so do the resource demands imposed on ecosystems and the impacts of our global footprint. Natural resources are not invulnerable and infinitely available. The environmental impacts of anthropogenic actions, which are processes or materials derived from human activities, are becoming more apparent – air and water quality are increasingly compromised, oceans are being overfished, pests and diseases are extending beyond their historical boundaries, and deforestation is exacerbating flooding downstream. It has been reported that approximately 40-50% of Earth's ice-free land surface has been heavily transformed or degraded by anthropogenic activities, 66% of marine fisheries are either overexploited or at their limit, atmospheric CO_2 has increased more than 30% since the advent of industrialization, and nearly 25% of Earth's bird species have gone extinct in the last two thousand years. Society is increasingly becoming aware that ecosystem services are not only limited, but also that they are threatened by human activities. The need to better consider long-term ecosystem health and its role in enabling human habitation and economic activity is urgent. To help inform decision-makers, many ecosystem services are being assigned economic values, often based on the cost of replacement with anthropogenic alternatives. The ongoing challenge of prescribing economic value to nature, for example through biodiversity banking, is prompting transdisciplinary shifts in how we recognize and manage the environment, social responsibility, business opportunities, and our future as a species.

The simple notion of human dependence on Earth's ecosystems probably reaches to the start of our species' existence, when we benefited from the products of nature to nourish our bodies and for shelter from harsh climates.

Recognition of how ecosystems could provide more complex services to mankind date back to at least Plato (c. 400 BC) who understood that deforestation could lead to soil erosion and the drying of springs. However, modern ideas of ecosystem services probably began with Marsh in 1864 when he challenged the idea that Earth's natural resources are not infinite by pointing out changes in soil fertility in the Mediterranean. Unfortunately, his observations and cautions passed largely unnoticed at the time and it was not until the late 1940s that society's attention was again brought to the matter. During this era, three key authors – Osborn, Vogt, and Leopold – awakened and promoted recognition of human dependence on the environment with the idea of 'natural capital'. In 1956, Sears drew attention to the critical role of the ecosystem in processing wastes and recycling nutrients. An environmental science textbook called attention to "the most subtle and dangerous threat to man's existence… the potential destruction, by man's own activities, of those ecological systems upon which the very existence of the human species depends". The term 'environmental services' was finally introduced in a report of the *Study of Critical Environmental Problems,* which listed services including insect pollination, fisheries, climate regulation and flood control. In following years, variations of the term were used, but eventually 'ecosystem services' became the standard in scientific literature.

Modern expansions of the ecosystem services concept include socio-economic and conservation objectives, which are discussed below. For a more complete history of the concepts and terminology of ecosystem services.

Examples

Detritivores like this dung beetle help to turn animal wastes into organic material that can be reused by primary producers.

Experts currently recognize four categories of ecosystem services. The following lists represent samples of each:

- food (including seafood and game), crops, wild foods, and spices
- water
- pharmaceuticals, biochemicals, and industrial products
- energy (hydropower, biomass fuels).

Regulating Services

- carbon sequestration and climate regulation
- waste decomposition and detoxification
- purification of water and air

- crop pollination
- pest and disease control.

Supporting Services

- nutrient dispersal and cycling
- seed dispersal
- Primary production.

Cultural Services

- cultural, intellectual and spiritual inspiration
- recreational experiences (including ecotourism)
- scientific discovery.

To understand the relationships between humans and natural ecosystems through the services derived from them, consider the following cases:

- In New York City, where the quality of drinking water had fallen below standards required by the U.S. Environmental Protection Agency (EPA), authorities opted to restore the polluted Catskill Watershed that had previously provided the city with the ecosystem service of water purification. Once the input of sewage and pesticides to the watershed area was reduced, natural abiotic processes such as soil adsorption and filtration of chemicals, together with biotic recycling via root systems and soil microorganisms, water quality improved to levels that met government standards. The cost of this investment in natural capital was estimated between $1-1.5 billion, which contrasted dramatically with the estimated $6-8 billion cost of constructing a water filtration plant plus the $300 million annual running costs.
- Pollination of crops by bees is required for 15-30% of U.S. food production; most large-scale farmers import non-native honey bees to provide this service. One study reports that in California's agricultural region, it was found that wild bees alone could provide partial or complete pollination services or enhance the services provided by honey bees through behavioural interactions. However, intensified agricultural practices can quickly erode pollination services through the loss of species and those remaining are unable to compensate for the difference. The results of this study also indicate that the proportion of chaparral and oak-woodland habitat available for wild bees within 1-2 km of a farm can strongly stabilize and enhance the provision of pollination services, thereby providing a potential insurance policy for farmers of this region.

- In watersheds of the Yangtze River (China), spatial models for water flow through different forest habitats were created to determine potential contributions for hydroelectric power in the region. By quantifying the relative value of ecological parameters (vegetation-soil-slope complexes), researchers were able to estimate the annual economic benefit of maintaining forests in the watershed for power services to be 2.2 times that if it were harvested once for timber.
- In the 1980s, mineral water company Vittel (now a brand of Nestlé Waters) faced a critical problem. Nitrates and pesticides were entering the company's springs in northeastern France. Local farmers had intensified agricultural practices and cleared native vegetation that previously had filtered water before it seeped into the aquifer used by Vittel. This contamination threatened the company's right to use the "natural mineral water" label under French law. In response to this business risk, Vittel developed an incentive package for farmers to improve their agricultural practices and consequently reduce water pollution that had affected Vittel's product. For example, Vittel provided subsidies and free technical assistance to farmers in exchange for farmers' agreement to enhance pasture management, reforest catchments, and reduce the use of agrochemicals. This is an example of a Payment for ecosystem services program.

Ecology

Understanding of ecosystem services requires a strong foundation in ecology, which describes the underlying principles and interactions of organisms and the environment. Since the scales at which these entities interact can vary from microbes to landscapes, milliseconds to millions of years, one of the greatest remaining challenges is the descriptive characterization of energy and material flow between them. For example, the area of a forest floor, the detritus upon it, the microorganisms in the soil and characteristics of the soil itself will all contribute to the abilities of that forest for providing ecosystem services like carbon sequestration, water purification, and erosion prevention to other areas within the watershed. Note that it is often possible for multiple services to be bundled together and when benefits of targeted objectives are secured, there may also be ancillary benefits – the same forest may provide habitat for other organisms as well as human recreation, which are also ecosystem services.

The complexity of Earth's ecosystems poses a challenge for scientists as they try to understand how relationships are interwoven among organisms, processes and their surroundings. As it relates to human ecology, a suggested research agenda for the study of ecosystem services includes the following steps:

1. identification of *ecosystem service providers* (*ESP*s) – species or populations that provide specific ecosystem services – and characterization of their functional roles and relationships;
2. determination of community structure aspects that influence how ESPs function in their natural landscape, such as compensatory responses that stabilize function and non-random extinction sequences which can erode it;
3. assessment of key environmental (abiotic) factors influencing the provision of services;
4. measurement of the spatial and temporal scales ESPs and their services operate on.

Recently, a technique has been developed to improve and standardize the evaluation of ESP functionality by quantifying the relative importance of different species in terms of their efficiency and abundance. Such parameters provide indications of how species respond to changes in the environment (i.e. predators, resource availability, climate) and are useful for identifying species that are disproportionately important at providing ecosystem services. However, a critical drawback is that the technique does not account for the effects of interactions, which are often both complex and fundamental in maintaining an ecosystem and can involve species that are not readily detected as a priority. Even so, estimating the functional structure of an ecosystem and combining it with information about individual species traits can help us understand the resilience of an ecosystem amidst environmental change.

Many ecologists also believe that the provision of ecosystem services can be stabilized with biodiversity. Increasing biodiversity also benefits the variety of ecosystem services available to society. Understanding the relationship between biodiversity and an ecosystem's stability is essential to the management of natural resources and their services.

The Redundancy Hypothesis

The concept of ecological redundancy is sometimes referred to as *functional compensation* and assumes that more than one species performs a given role within an ecosystem. More specifically, it is characterized by a particular species increasing its efficiency at providing a service when conditions are stressed in order to maintain aggregate stability in the ecosystem. However, such increased dependence on a compensating species places additional stress on the ecosystem and often enhances its susceptibility to subsequent disturbance. The redundancy hypothesis can be summarized as "species redundancy enhances ecosystem resilience".

The Rivet Hypothesis

Another idea uses the analogy of rivets in an airplane wing to compare the exponential effect the loss of each species will have on the function of an ecosystem; this is sometimes referred to as *rivet popping*. If only one species disappears, the efficiency of the ecosystem as a whole is relatively small; however if several species are lost, the system essentially collapses as an airplane wing would, were it to lose too many rivets.

The hypothesis assumes that species are relatively specialized in their roles and that their ability to compensate for one another is less than in the redundancy hypothesis. As a result, the loss of any species is critical to the performance of the ecosystem. The key difference is the rate at which the loss of species affects total ecosystem function.

The Portfolio Effect

A third explanation, known as the *portfolio effect,* compares biodiversity to stock holdings, where diversification minimizes the volatility of the investment, or in this case, the risk in stability of ecosystem services. This is related to the idea of *response diversity* where a suite of species will exhibit differential responses to a given environmental perturbation and therefore when considered together, they create a stabilizing function that preserves the integrity of a service. Several experiments have tested these hypotheses in both the field and the lab. In ECOTRON, a laboratory in the UK where many of the biotic and abiotic factors of nature can be simulated, studies have focused on the effects of earthworms and symbiotic bacteria on plant roots. These laboratory experiments seem to favour the rivet hypothesis. However, a study on grasslands at Cedar Creek Reserve in Minnesota seems to support the redundancy hypothesis, as have many other field studies.

Economics

There is extensive disagreement regarding the environmental and economic values of ecosystem services. Some people may be unaware of the environment in general and humanity's interrelatedness with the natural environment, which may cause misconceptions. Although environmental awareness is rapidly improving in our contemporary world, ecosystem capital and its flow are still poorly understood, threats continue to impose, and we suffer from the so-called 'tragedy of the commons'.

Many efforts to inform decision-makers of current versus future costs and benefits now involve organizing and translating scientific knowledge to economics, which articulate the consequences of our choices in comparable units of impact on human well-being. An especially challenging aspect of this process is that interpreting ecological information collected from one spatial-temporal scale does not necessarily mean it can be applied at another; understanding the dynamics of ecological processes relative to ecosystem services is essential in aiding economic decisions. Weighting factors such as a service's irreplaceability or bundled services can also allocate economic value such that goal attainment becomes more efficient.

Wetlands can be used to assimilate wastes. The economic valuation of ecosystem services also involves social communication and information, areas that remain particularly challenging and are the focus of many researchers. In general, the idea is that although individuals make decisions for any variety of reasons, trends reveal the aggregative preferences of a society, from which the economic value of services can be inferred and assigned. The six major methods for valuing ecosystem services in monetary terms are:

Avoided Cost

Services allow society to avoid costs that would have been incurred in the absence of those services (e.g. waste treatment by wetland habitats avoids health costs)

Replacement Cost

Services could be replaced with man-made systems (e.g. restoration of the Catskill Watershed cost less than the construction of a water purification plant)

Factor Income

Services provide for the enhancement of incomes (e.g. improved water quality increases the commercial take of a fishery and improves the income of fishers)

Travel Cost

Service demand may require travel, whose costs can reflect the implied value of the service (e.g. value of ecotourism experience is at least what a visitor is willing to pay to get there)

Hedonic Pricing

Service demand may be reflected in the prices people will pay for associated goods (e.g. coastal housing prices exceed that of inland homes)

Contingent Valuation

Service demand may be elicited by posing hypothetical scenarios that involve some valuation of alternatives (e.g. visitors willing to pay for increased access to national parks)

Economic Opportunity

Treating the environment as an externality may generate short-term profit at the expense of sustainability. Sustainable business practices, on the other hand, integrate ecological concerns with social and economic ones (i.e., the triple bottom line). Growth that depletes ecosystem services is sometimes termed "uneconomic growth" as it leads to a decline in quality of life. Minimising such growth can provide opportunities for local businesses. For example, industrial waste can be treated as an "economic resource in the wrong place". The benefits of waste reduction include savings from disposal costs, fewer environmental penalties, and reduced liability insurance. This may lead to increased market share due to an improved public image. Energy efficiency can also increase profits by reducing costs.

The idea of sustainability as a business opportunity has led to the formation of organizations such as the Sustainability Consortium of the Society for Organizational Learning, the Sustainable Business Institute, and the World Council for Sustainable Development. The expansion of sustainable business opportunities can contribute to job creation through the introduction of green-collar workers.

THE ECONOMIC PERSPECTIVE OF SUSTAINABLE DEVELOPMENT

From the point of view of neoclassical economic theory, sustainability can be defined in terms of the maximization of welfare over time. (This is assumed to be human welfare – we will introduce the claims of the non-human world when we consider the ecological perspective.) Most economists simplify further by identifying the maximization of welfare with the maximization of utility derived from consumption. While this may be criticized as an oversimplification, it certainly includes many important elements of human welfare (food, clothing, housing, transportation, health and education services, etc.) and it has the analytical advantage of reducing the problem to a measurable single-dimensional indicator.

A formal economic analysis then raises the question of whether sustainability has any validity as an economic concept. According to standard economic theory, efficient resource allocation should have the effect of maximizing utility from consumption. If we accept the use of time discounting as a method of comparing the economic values of consumption in different time periods, then sustainability appears to mean nothing more than efficient resource allocation – a concept already well established in economics.

One line of criticism of this reductionist approach to sustainability centres on the use of discounting. At a discount rate of 10%, the value of $1 million one hundred years from now is the same as a mere $72 today. Thus it would apparently be justifiable to impose costs of up to $1 million on people in the year 2100 in order to enjoy $72 worth of consumption today. By this logic, much resource depletion and environmental damage could be considered acceptable, and even optimal, according to a criterion of economic efficiency.

The problem is that in accepting the use of a discount rate, we have implicitly imposed a specific choice regarding the relative welfare of present and future generations. Howarth and Norgaard have shown that the choice of a discount rate is equivalent to a choice of allocations among generations. Use of a current market discount rate gives undue weight to the preferences of current consumers. When we consider issues such as soil erosion or atmospheric buildup of greenhouse gases, where the most damaging impacts are felt over decades or generations, this creates a strong bias against sustainability. Thus to achieve intergenerational equity, we must either impose a low discount rate or some kind of sustainability rule regarding resource use and environmental impacts. A related issue concerns the concept of natural capital. Soils and atmospheric functions are aspects of natural capital, which consists of all the natural resources and environmental services of the planet. Herman Daly has suggested that sustainable development can be operationalized in terms of the conservation of natural capital. This policy goal leads to two decision rules, one for renewable and the other for non-renewable resources.

For renewables, the rule is to limit resource consumption to sustainable yield levels; for non-renewables the rule is to re-invest the proceeds from non-renewable resource exploitation into investment in renewable natural capital. Following these two rules will maintain a constant stock of natural capital. To maintain a constant *per capita* stock of natural capital also requires a stable level of human population, a factor which Daly has also emphasized.

This suggestion of a specific sustainability decision rule for natural capital is quite different from the standard neo-classical approach. In the neo-classical view, there is no special reason to conserve natural capital. A well-known principle derived from work by Solow and Hartwick (the "Hartwick rule") states that consumption may remain constant, or increase, with declining non-renewable resources provided that the rents from these resources are reinvested in reproducible capital. Unlike Daly's reinvestment rule, this does not require maintenance of any particular stock of *natural* capital.

The essential assumption involved in the Solow/Hartwick approach is that of *substitutability* of the two types of capital. If, for example, we cut down forests but build factories, we are better off provided the economic value of the new industrial plant exceeds the economic value of the lost forests. Daly's view is based on the opposite assumption, that "man-made and natural capital are fundamentally complements and only marginally substitutes." If natural capital has a special and unique importance, then neo-classical economic efficiency will not suffice for sustainability.

Michael Toman has suggested that the issue may be resolved by recognizing that some issues can be appropriately dealt with through neo-classical market efficiency, while others require the application of a "safe minimum standard" approach to protect essential resources and environmental functions. He suggests that the criteria of possible severity and irreversibility of ecological damages should be used to decide which theoretical framework is more appropriate:

The concept of a safe minimum standard can be applied to concerns about intergenerational fairness, resource constraints, and human impact. The safe minimum standard posits a socially determined, albeit "fuzzy," dividing line between moral imperatives to preserve and enhance natural resource systems and the free play of resource trade-offs... Following a safe minimum standard, society would rule out actions that could result in natural impacts beyond a certain threshold of cost and irreversibility. Central to the safe minimum standards approach are the role of public decision making and the formation of societal values.

The adoption of this reasonable suggestion would have far-reaching implications for economic theory and policy. Note the essential role of "moral imperatives," "public decision making," and "the formation of social values" in Toman's suggested decision framework. None of these appear in the neo-classical economic model, where markets are presumed to be the best resource allocators, and the occasional correction of a "market imperfection" the only appropriate role for government. Thus Toman is in effect asserting the importance of sustainability as a concept independent of standard neo-classical economic analysis, one which requires an explicitly normative and socially determined process of decision-making.

This represents a fundamental shift in the economic paradigm. Much as the Keynesian revolution validated the concept of government intervention to achieve macroeconomic stability, the acceptance of sustainability as a valid social goal places a new complexion on all policy issues concerning the relationship between human economic activity and the environment. Markets may be valuable and essential means, but they cannot determine the ends, which must be arrived at by a social decision process informed by different disciplinary viewpoints. This will require an unaccustomed humility on the part of economists, and a willingness to work together with other social and natural scientists. As Toman suggests:

There is great scope for interdisciplinary work to address some key issues related to sustainability, including defining objectives, identifying constraints, and resolving the relevant disagreements. Economists could make greater use of ecological information and the implications of physical resource limits in an analysis of resource values. Social scientists can contribute to an understanding of how future generations might value different attributes of natural environments. Ecologists should provide ecological information in a manner that can be used in economic valuation. They should also take into account the role of economic incentives in ecological impact analysis.

In order to explore further the implications of this approach, we need to examine the ecological and social dimensions of the issue. Then we can return to the question of whether a new paradigm for development policy has truly emerged from the multidisciplinary discussion on the nature of sustainability.

THE ECOLOGICAL PERSPECTIVE OF SUSTAINABLE DEVELOPMENT

Unlike economists, whose models provide no upper bound on economic growth, physical scientists and ecologists are accustomed to the idea of limits. Natural systems must exist subject to the unyielding laws of thermodynamics, and the science of population ecology has explored the implications of these laws for living organisms. As ecologist C.S. Holling puts it: Two of the fundamental axioms of ecological and evolutionary biology are that organisms are exuberantly over-productive, and that limits set by time, space, and energy are inevitably encountered. The foundations for all modern ecology and evolutionary biology rest in part upon the consequences of these two axioms. In an ecological perspective, then, sustainability must involve limits on population and consumption levels.

These limits apply to all biological systems. While humans may appear to evade them for a time, they must ultimately accept the boundaries of a finite planet. Ecologist Paul Ehrlich and colleagues have estimated that humans are now "consuming, co-opting, or eliminating some 40% of the basic energy supply for all terrestrial animals." Clearly, a doubling of this demand, as might well be implied by a 33% growth in population (to 8 billion) and a 50% growth in per capita consumption by 2050, would leave little room for any other species on the planet.

However, this simple assertion of limits does not fully capture the contribution of ecologists to the discussion of sustainability. What C.S. Holling identifies as a third axiom of ecology has even more significant implications. The third axiom "concerns processes that generate variability and novelty" – the generation of genetic diversity and the resultant processes of evolution and change in species and ecosystems.

Genetic diversity gives rise to *resilience* in ecosystems. Resilience is a "bounce-back" capacity which enables a system to respond to disturbances or damage. For example, a forest ecosystem may recover from a pest infestation through an increase in the population of predators which control the pest, an expansion of species unaffected by the pest, and possibly a development of pest resistance in affected species.

The patterns of response will be widely variable, but the essential integrity of the ecosystem will be preserved. The key to resilience is the existence of a wide variety of species, interacting with each other and providing a reservoir of genetic forms which provide the potential to adapt to changing conditions.

For the ecologist, then, sustainability should be defined in terms of the maintenance of ecosystem resilience. This view of sustainability is clearly different from the human-centred conceptions put forward by the World Commission on Environment and Development and the consumption-based principles proposed by economic theorists. This contrast has been explored by Common and Perrings, who distinguish between "Solow-sustainability", derived from the economic model of stable or increasing consumption, and "Holling-sustainability", based on ecosystem resilience. They find that "the concepts of Solow-sustainability and Holling-sustainability are largely disjoint. This implies that there may be no close relationship between economic efficiency and ecological sustainability."

The importance of the ecological perspective is increasingly evident, as more of the critical problems facing humanity arise from failures of ecological resilience. The resurgence of diseases due to the development of antibiotic resistance, the disruption of ecosystems by introduced species, the formation of "dead zones" in coastal waters, and the multiple ecological threats related to climate change and increased climate volatility, all testify to the impacts of expanding human economic activity.

As Holling puts it: Increasing human populations in the South, and the planetary expansion of their influence, combined with exploitative management in both North and South, reduces functional diversity and increases spatial homogeneity not only in regions but on the whole planet.

Functional diversity of the structuring processes and spatial heterogeneity are the two most critical determinants of ecological robustness and resilience, the attributes that provide the reserve of ecological services and of time that have allowed people to adapt and learn in the past. And now these critical attributes are being compromised at the level of the planet.

The horrifying impact of AIDS, most especially on the African continent, is perhaps the worst example to date of the feedback effects of human destruction of ecosystem resilience. AIDS probably originated in rain-forest primates, and spread to humans through human intrusion into the forest. Rather than remaining isolated in small communities, it then spread worldwide through global commerce and travel, like many other destructive viruses and pests. Population checks through such drastic ecological backlash are, of course, familiar to ecologists. But they are generally far from the thoughts of the economists and policymakers who up until now have shaped our conceptions of development.

Sustainability, then, is more than limits on population or restraint in consumption – though these are important. It means that in our choice of goods and technologies we must be oriented to the requirements of ecosystem integrity and species diversity.

It also implies that the apparent independence of economics from biophysical science is a luxury we can no longer afford. Common and Perrings suggest that: An ecological economic approach requires that resources be allocated in such a fashion that they threaten neither the system as a whole nor the key components of the system. For the system to be sustainable it must serve consumption and production objectives that are themselves sustainable. If existing preferences and technologies, as perpetuated and sanctified in the concept of consumer sovereignty, are not sustainable, then the system as a whole will be unstable. The appropriate policy instruments to address these concerns are varied and complex.... What is important is that an ecologicaleconomics of sustainability privileges the needs of the system over those of individuals.

Clearly, an integration of economics and ecology is required, and this can only be achieved with the assistance of the third element of the sustainability triad – the social perspective. If we cannot rely on unregulated markets to solve our problems, we must turn to conscious social action. But social action by whom, and at what level? And how do the environmental issues relate to the other great failure of development to date – the persistence of inequality? It is in the social area that we must seek the key to the formulation of policies for sustainable development.

SOCIAL PERSPECTIVE OF SUSTAINABLE DEVELOPMENT

Advocates of sustainable development, as we have noted, recognize the social component of development as an essential part of the new paradigm. In doing so, they are validating the importance of a much older perspective. A "human development" approach emphasizing issues of basic needs and equity is well grounded in the history of economic theory. Sudhir Andand and Amartya Sen point out that concerns for these dimensions of economic development start with the earliest economic theorists, and contrast the human development approach to the wealth maximization approach which has dominated modern economics:

There is... no foundational departure in making economic analysis and policy take extensive note of the demands of human development. This approach reclaims an existing heritage, rather than importing or implanting a new diversion.... The interest in human development has had to compete with other priorities and pursuits within the body of mainstream economics. The preoccupation with commodity production, opulence, and financial success can also be traced in economics through several centuries...Indeed, the dominant contemporary tradition of focusing on such variables as per capita gross national product (GNP) or national wealth is a continuation – perhaps even an intensification – of the old opulence-oriented approach.

As we have noted, the focus on basic needs and equity in development has been represented by the United Nations Development Programme's series of Human Development Reports. In addition to calculating the Human Development Index which offers a different measure of development success from per capita GNP or GDP27, the Human Development Reports focus each year on a different aspect of social and economic development, such as democratic governance (1993), gender inequity (1995), and poverty (1997).

The HDI combines life expectancy, adult literacy, and school enrollment ratios with per capita GDP in a weighted average to get an index between 0 and 1. The results clearly show that development is a multidimensional process, and that higher GDP does not necessarily mean higher overall welfare.

Some countries, such as Costa Rica (HDI = 0.883) and Sri Lanka (HDI= 0.704), stand out in terms of their human development well above others of almost identical GDP per capita (Brazil and Turkey, comparable to Costa Rica in per capita GDP, have HDI = 0.804 and 0.792 respectively; Congo and Pakistan, almost identical in GDP to Sri Lanka, have HDI's of 0.538 and 0.483 respectively).

While the HDI does not explicitly include any environmental measures, the 1994 report discussed the relationship between sustainability and equity: The concept of sustainable development raises the issue of whether present life-styles are acceptable and whether there is any reason to pass them on to the next generation.

Because intergenerational equity must go hand in hand with intragenerational equity, a major restructuring of the world's income and consumption patterns may be a necessary precondition for any viable strategy of sustainable development... Development patterns that perpetuate today's inequities are neither sustainable nor worth sustaining. In the 1997 report, a section on "Resisting New Forces of Poverty" discusses factors which cause worsening conditions for the world's poor. Prominent among these are the HIV/AIDS pandemic which is "creating a new wave of impoverishment – and reversing earlier gains." (We have already noted the relationship of AIDS and resurgent diseases to ecological degradation.) Another factor is environmental degradation on marginal lands – the dry, swampy, saline, and steep areas where many of the rural poor struggle to survive.

Clearly, the issue of environmental sustainability is intertwined with that of poverty and inequity. It has frequently been noted that the causative relationship runs both ways – increased poverty and loss of rural livelihoods accelerates environmental degradation as displaced people put greater pressure on forests, fisheries, and marginal lands.

If the problems of environment and equity are clearly related, then so must be the solutions. Third World critics of the standard, "Western" development model see that model itself as a significant cause of the problems. The sweeping optimism implicit in Rostow's original stages-of-growth paradigm ignores social and cultural differences between nations, as well as the fundamental power disparity between developed and developing nations. A view of development as a one-way journey to improved conditions fails to match the experience of many people whose livelihoods are threatened by globalization.

Local experiences of western development in many localities of the third world have been closely associated with the dissolution of indigenous cultural, political, and economic systems; with increased inequalities in life chances between genders and among classes, castes, and ethnic groups; and with deterioration in, and removal of access to, the biophysical environment.

As we seek for models of sustainable and equitable development, then, we must recognize the need for a fundamental revision in what Pablo Escobar refers to as a "development discourse" dominated by the power and "modernizing" vision of the West. A sustainable development process will have to be democratized, decentralized, and pluralistic.

It will have to balance wealth-creation with wealth distribution. And it will have to include a healthy skepticism about Western models and the modernizing effects of global markets.

It is not only radical critics who are aware of the need for significant changes in the development paradigm. The World Bank has recently produced reports stressing the importance of social capital, the role of the state, and the importance of local government and non-governmental organizations in development.

From the Bank's more conventional perspective, participatory democracy, decentralization, and social capital represented by strong local organization, are compatible with, and beneficial to, standard measures of development such as GDP per capita. However, simply highlighting the importance of these factors is a new departure for market-oriented economic theorists. In addition, the World Bank has produced research on indicators of sustainable development, in particular measures of *genuine savings:* "the true rate of savings in a nation after due account is taken of the depletion of natural resources and the damages caused by pollution". This new attention to a combination of social and environmental factors indicates that lines of thought formerly at the fringes of development policy are making their way into the mainstream.

What has been referred to as the "Washington consensus" on the virtues of free markets and globalization has also come under challenge from the World Bank's own chief economist, Joseph Stiglitz. Stiglitz argues that there are many areas in which the operations of "free markets" are flawed by asymmetric control of information. This rather abstruse economic theory can be translated into an awareness of the importance of institutions and social norms in shaping market outcomes. This in turn justifies social and governmental action at both the micro and macro levels, and opens the way to a more explicitly normative theory of development. In this sense, Stiglitz is returning to the more goal-oriented perspective of the of the original theorists of development – except that the goals which now seem appropriate have much stronger social and environmental components. While there are clearly wide differences of perspective and emphasis between the critics within and without the development establishment, there seems to be a widely felt discontent with present development theory and practice, and it appears that the elements of a new paradigm are emerging. Can we combine the economic, ecological, and social perspectives to provide a new vision of development in the twenty-first century?

3

NGOs, Democracy and Sustainable Development

NGOs have rapidly gained an international reputation. They are being consulted on matters of policy making, planning and implementation at the local, national and international levels. This is because they offer new perspectives as well as a wealth of experience in such areas as grassroots development, environmental protection and the defence of human rights.

They have a broad knowledge base and strong commitment to issues relevant to the emancipation of the poor. They have shown more sensitivity and understanding of Africa's severe economic and political crisis than their respective governments and the multilateral institutions.

Keeping in mind that Africans at the grassroots level must have the lead in defining their needs and formulating development strategies, NGOs have recommended an action-orientated 'compact' for African development which must be translated into coordinated programmes for long-term solutions. Indeed, many NGOs are more active and knowledgeable in the area of development needs of poor countries than are their governments. They have demonstrated an awareness of the fact that conventional development models have not changed the situation of the poorest in Africa and they have been quick to recognize that people are poor because they have no power.

NGOs are increasingly influencing their governments, mobilizing people at the grassroots level by strengthening their institutions and raising their awareness; bringing to the attention of donor countries and the multilateral institutions the harmful effects of some of their policies; calling for equity-led strategies that give priority to achieving broad participatory ownership, control and management of natural resources by people to serve their own needs; and decrying current development strategies which favour the market economy and lead to more debts and reliance on exports and cash crops, whose effects on environmental degradation and overuse of agricultural land are apparent.

NGOs are increasingly interested in sustainability and they realize that local development initiatives will be sustainable only when in partnership with a supportive national development system. That is why they are paying more attention to local and national governments which control the resources and policies that have a bearing on grassroots development, in the hope of influencing them in favour of an institutional and policy setting that is supportive of sustainable development policies. This task requires NGOs to have intra-and inter-organizational skills as well as technical competence and close collaboration with each other.

NGOs now see the future development of African countries as achievable through broader participation in the decision making process. They see such participation as informing both national and local development decisions. In conferences, NGOs have made it clear that the absence of full democratic rights in Africa is the main cause of Africa's economic decline. That is why they are advocating human centred development, the democratization of the development process, and are taking the lead in advocating policy and institutional reforms which are supportive of sustainable development and democracy and would stop the downward spiral of economic inequity and ecological destruction.

NGOs have identified the pressing problems facing African countries as being the undemocratic systems of government and the unequal, unsustainable, misguided and inappropriate development strategies. Consequently, they have proposed strategies by which equitable distribution of the benefits of sustainable development could be achieved; these include reforming of the world's trading system, more financial resources to African countries, and a reduction in military expenditure. They are critical of the institutions and policies causing environmental, economic and social degradation. For example, transnational corporations (TNCs) are under attack because of their role in the extraction of natural resources and despoliation of the environment, and international financial institutions are decried for opening up African economies for resource extraction by TNCs. For NGOs to be able to promote democracy and sustainable development, they are recommended to "network; build up relationships with governments; participate in development planning committees at regional, provincial, district and ward levels; achieve economic self-reliance and mobilize, build up and expand constituencies," otherwise their democratizing potential will be limited. The existence of a dense network of autonomous grassroots development organizations and African NGOs on a substantial scale is essential in order to exert pressure in the interests of the poor and to act as their representatives, as well as a countervailing force to the power of the state, bureaucrats and local elites. NGOs also need to work to strengthen the organizational, technical and managerial capabilities of grassroots organizations so that the latter are able to stand up and press for demands and hold governments accountable for their actions.

By strengthening and developing grassroots organizations' capacities, NGOs will be enabling them to ensure resource management and control which correspond to the specific local context. Considering the climate in which NGOs must operate, their position on the African scene can sometimes be unenviable. NGOs have to contend with policy, economic and above all, political instability with all their ramifications. In some countries, the rise of political pluralism and the decline of centralized, one-party regimes has been replaced by fragmented special interests which have produced polarization, violence and political paralysis. As if this were not enough, mutual suspicion and hostility sometimes characterize government-NGO relationships. A confrontational relationship with governments does little to enable NGOs go about their tasks effectively.

NGOs may face other threats, too, including disbanding or control. Other permanent threats to NGOs are administrative cooptation, appropriation, harassment and politically-motivated legislation. NGOs have also to contend with vested interests, which include bureaucrats, politicians and rural power elites, many of whom will oppose attempts to transfer power, responsibilities and resources to local institutions. Vested interests which have gained from non-democratic regimes are likely to oppose political and economic reforms. In such circumstances, governments, as well as vested interests, can constrain NGOs' ability to promote democracy and sustainable development. One way to overcome this is for NGOs to strive to create political space, build coalitions of friends and identify common points with governments, and then use this strength to chip away at the power of the vested interests.

Balance Sheet

The major sources of funding for NGOs continue to be voluntary private sources and governments. Where before African governments were the only major recipients of official aid from donor countries and multilateral institutions, nowadays more aid is being channelled through NGOs, many of which are perceived by the donor community to work more efficiently in participatory development and to operate in those areas which are not accessible to governments. The effect of this re-direction of some of the aid hitherto going to African governments has meant that these governments are being bypassed as implementing agents. This is leading to interesting examples of how governments are trying to 'coordinate' NGOs in the hope that they might be able to 'control' them now that they are recipients of more official funding.

While increasing availability of public funds for NGOs has been welcomed, for it expands their operations, NGOs are concerned that increasing acceptance of such funds could compromise their development goals, with the risk that they will be increasingly seen as agents of governments and multilateral institutions rather than as partners in development. The central challenge facing NGOs is how to maintain their voluntary character while becoming increasingly effective in their work.

NGOs realize that dependence on others for funding may compromise their flexibility to deal with pressing development issues. It may also undermine their ability to speak out against those policies of funders which they see as harmful to the interests of the poor. The other quandary which NGOs face is that while they advocate sustainable development and democracy for those they support, they themselves are, in many instances, neither democratic nor self-reliant.

In highlighting the potential and actual impact of NGOs as promoters of sustainable development, it is important that we do not lose sight of the difficulties NGOs face in achieving their goals. A recent series of studies by the UK's Overseas Development Institute (ODI) in Zimbabwe, India and Bangladesh suggests that NGOs have not been successful in (a) benefiting the poorest households, (b) benefiting women, and (c) ensuring self-sustainability of local NGOs. These findings serve to remind us that the tasks that NGOs set themselves are not easy to achieve and that a lot more will have to be done within and outside the NGOs if they are to be able to promote development that is balanced between local short-term use of man's environment and the maintenance and enhancement of long-term productivity.

NGOs have yet to make a significant impact both at national and local levels. In part, this inability stems from their lack of leverage. At the local level, they have yet to develop into an effective force which could counter the presence of local elites and influence decisively people's attitudes as well as local development policies. NGOs have many constituents, namely, donors, beneficiaries, policy makers and the public. These constituents have conflicting expectations. NGOs need to develop mechanisms for multiple accountabilities.

Currently such mechanisms do not exist. This serves to constrain NGOs' effectiveness and blunts any claim to democratic credentials.

The political stance of many African NGOs contrasts with that of their Latin American counterparts where NGOs go as far as providing "trained personnel to occupy government positions." Many Latin American NGOs have had a history of directly and openly supporting social movements in opposition to military regimes. NGOs in Africa have not taken that high a political profile. For many African NGOs, politics is a forbidden and dangerous area. They argue that the best way for them to fulfil their development roles is to remain apolitical. Yet this is an unrealistic strategy, not least because many NGOs' operational activities are themselves political in that they seek to shift existing inequities in favour of the poor. Instead, they prefer to promote democracy by supporting grassroots organizations with resources, training and information. By their own actions, especially in participatory projects, NGOs have created practical schools of democracy from which members from grassroots organizations are able to follow democratic practices. Nevertheless, the feeling is that NGOs in Africa should do much more to contribute towards a more democratic Africa.

The immense development challenges posed by Africa cannot find all the solutions in NGOs. NGOs have great capacity but cannot be the panacea to the challenges. That said, for NGOs' potential and impact to materialize fully, they must have more effective systems of internal monitoring and self-evaluation and must achieve self-sustainability. Somehow, they must acquire the political skills which will see them through the complex socioeconomic and political maze of the African situation. Strategically, close collaboration with national governments is essential and, from time to time, NGOs must not hesitate to appeal to multilateral donors who have leverage and could act to protect them against political interference.

The ability of NGOs to promote democracy and sustainable development policies in Africa should be considered in the context of the states' preparedness to accept reforms which permit "devolution of power and responsibility for resource use and management from the centre to the communities." The goodwill of national governments, as well as of the international community, is crucial to NGOs' successes or failures, not least because, in the case of the latter, they have the resources to create a supportive international climate that could reverse the flow of resources from African to northern countries and support better terms of trade and accessible markets.

As for national governments, they should show goodwill by creating internal conditions which permit a liberal democratic framework as well as a supportive legal framework. Both macro-and micro-economic policies which are favourable to NGOs would offer a conducive framework within which NGOs would flourish. In their quest for sustainable development and a democratic society, therefore, NGOs must strive to attain this goodwill. It underlies all else.

ROLE OF NGOS IN SUSTAINABLE RURAL DEVELOPMENT

20th century transferred the number of issues to its successor with regard to social sector. Perhaps among those most discussion able and dispersing one is "The Role of NGO, s in the Development" particularly with regard to third world having mushroom growth largely depending upon the contribution made by the socially developed countries, which on its turn also exalted a number of issues lying under the generous contribution made, by the developed world.

But as far as our concern, the presentation contains in itself the evolutionary development of the concept both theoretically and practically emerging the various types of NGOs i.e. charitable organizations, national organizations, community organization boards and international NGOs etc. The objectives of these NGOs as relief welfare, community development, sustainable system and people's participation are also considered in this work. The role of NGOs which are common for almost all the Ngo, s with regard to development in different fields as in education, health, women welfare etc are mostly sponsored by international NGOs. NGOs also contribute their due share in the development sector of Pakistan and the facts about their activities funds and utilization of funds is also considered in this presentation. The NGOs are also working in rural areas of Pakistan and their programs in these areas are also under the consideration of our paper so, all the presentation will provide the knowledge and basic facts about the role of NGOs in development both at national and international level, their types, work ability and objectives and some fact about their weaknesses.

INTRODUCTION

The term NGO seems to be deceptively simple. It may overlook the enormous variety and differential capabilities of different NGOs. In fact, NGOs offer a kaleidoscopic collection of organizations varying in origin, size, programmes, ideology, role strategy, funding, linkages evaluation, problem etc. NGOs embrace a bewildering group of organizations varying in terms of innumerable parameters. No standard definition can include all organizations working at present under the title of NGO, originally voluntarism was a doctrine which held that the will is dominant factor or it is a principle relying on one's own free will for an action. The definition of NGOs vary as:

1. According to Asian Development Bank the term non-governmental organization refers to organization
   - Not based in government.
   - Not created to earn profit.
2. United Nations defines it "NGOs are private organizations that pursue activities to relieve suffering, promote the interest of poor, protect the environment, provide basic social services or undertake community development"

BACKGROUND OF NGOS IN DEVELOPMENT

Although NGOs have recently emerged into the development limelight but they are not a recent phenomenon. They were the earliest form of human organizations. Long before the governments, people organized themselves into group for mutual protection and self help.

First, there were farmer's organizations as in Japan in 1868; such organizations played a vital role in agricultural movement. Traditional self-help associations have also a long history in Africa and Asia.

During the 18th and 17th centuries in particular there has been an explosion in the number of NGOs and an upsurge for the realistic answers to problem over a king of neglected issues related to ecological degradation, rights of people and other common property resources appropriate technologies, health, safety, gender and equity.

The institutional forms to such organizations can be traced back in late 19th and early 20th centuries particularly in west world where the history of social organizations seems to have been largely influenced by "laissez fair" movement based on a more planned way.

NEW TRENDS IN NGOS ACTIVITY (PEOPLE PARTICIPATION)

New trends emerge in NGOs activities from 1950 to 1960 when it start to work in field of development. Similarly, the concept of people's participation does not have a long history. It reflects partly the failure of the" trickle down" model of economic development advocated after World War II. In 1980,s NGOs become a major phenomenon in the field of development. Tvedt analyzed NGOs "as an outcome of complicated processes where factors like international ideological trends, donor policies and agenda interacts with national historical and cultural conditions in a complex way. On the whole these organizations are commanding growing attention as possible alternative to government in addressing the needs of vast of population. So, we can summarize NGOs development in three stages.

- Social and cultural in early stage.
- Community services and development in intermediate stage.
- More recently target oriented activist groups.

Need for NGOs

There is none the less a single answer to question why NGOs are formed? How they are given meaning and how they operate? One cannot perceive NGOs as entities but we have taken into account the notion of multiple relation. The entry of NGOs in the field of development process thus represents important response to the need resulted due to the overburdened government, the hesitant private sector and underutilized people power. These are appeared to compose of overlapping social networks.

The development experience of 1970s and 1980s have raised more and more critical concerning as growing awareness about the widening gap between very few rich and the vast majority of poor in developing countries. This has also given a momentum to search for a more adequate and appropriate strategy for improving conditions. So, strategies constitute basic elements of the development of a number of NGOs throughout the world, which get people's participation. recent, global transformations and the search to a variable new option for supporting grass-root development presently provide quite significant opportunities for a rapid development of NGOs in the decade of 1980s in following consideration:

I. Growing interest among donors and national governments in strengthening the development roles of institution outside the public sector.

II. The demonstrated capacity of some non-governmental organizations to reach the poor more effectively than public agencies.

III. A sharp decline in public development resources, necessating a search by government for more cost affective alternatives to conventional public services and development programs.

IV. Ability to carry out programme on national scale and influence national policies and agencies.

Today, the NGOs address every conceivable issue and they operate virtually in every part of the globe. Though international NGOs activity has grown steadily, most NGOs operate within a country and frequently they function properly. According to one estimate some 25000 NGOs now qualify as international NGOs up from less than 400 a century ago.

ROLES OF NGO ACCORDING TO THE EXPECTATION OF PEOPLE

NGOs play a critical role in all areas of development. People and policy makers are agree on one thing that NGOs play a very important role in development. Role of NGOs vary over the years as the policy of government changes. NGOs are almost dependent on polices of government. Socio economic development is a shared responsibility of both i.e. government and NGOs. Role of NGOs are complementary but vary according to polices of government.

If we closely pursue the voluminous literature on NGOs many roles can be found according to the expectations of people. The major development roles ascribed to NGOs are to act as:

- Planner and implementer of development programmers,
- Mobiliser of local resources and initiative,
- Catalyst, enabler and innovator,
- Builder of self reliant sustainable society,
- Mediator of people and government,
- Supporter and partner of government programme in activating delivery system implementing rural development programmes, etc.,
- Agents of information,
- Factor of improvement of the poor, and
- Facilitator of development education, training, professionalisation, etc.

Basically NGOs role is to prepare people for change. They empower the people to overcome psychological problem and opposition of oppress. Its role cannot be denied.

OBJECTIVES OF NGOS IN DEVELOPMENT

NGO is one of the alternatives available among various development organizations and one of the inputs among technical, financial and other resources, major merits of NGOs are emerging from their limited scale of operation; the sporadic efforts of NGOs can be consolidated and made more effective. Still the primary role of NGO is at the local level as mobilizes of people and their resources for an indigenous self-sustainable development. And at this level it can be a pioneer, mediator power broker, catalyst and has many other roles. NGOs and their long lomerations also are very in playing their role as advocates in policy issues beyond local level-national or even international level. Proper assessment of expected an actual roles of NGOs enable us to make them an effective alternative in the development process. However, small and sporadic NGOs are, they are valued in a pluralist society as an alternative approach to conventional system of attaining human well being and as such NGOs have a pivotal role to play in any society especially where institutions are alienated and development is dehumanized.

An NGO is nowadays not expected to deliver directly some benefits to people, but to motivate people, mobilize resources, initiate leadership, and participate in development programmes for self reliance. An NGO is only an enabler and as and when a society is made self reliant, role of NGO is shifted to another place where NGO service is required. But NGO works in relief and delivery of public goods as direct suppliers and majority of the development NGOs are also involved directly in productive activities. The roles of NGOs an enabler or catalyst for self reliant society and as supplier or implementer is relevant where bureaucracy is indifferent or inefficient, programs lack flexibility and cost effectiveness poor are ignorant, elite are ambitious, successes and services are pre conditions for motivation etc.

TYPES OF NGOS

Types of NGOs can be understand by their level of orientation and level of cooperation. Types of NGOs by the level of orientation. It has further types as under,

i. Charitable orientation. It often involves a paternalistic effort with little participation by "beneficiaries". It includes the NGOs which directed the people towards meeting the needs of poor and help them by gaining them food, clothing, medicine, provision of housing etc. such NGOs may also undertake relief activities during natural or man made herds.

ii. Service orientation. It includes with NGOs with activities such as the provision of health, family planning or education services in which the program is designed by the NGOs and people are expected to participate in its implementation and in receiving the services.

iii. Participatory orientation. It is characterized by self-help projects where local people are involved particularly for example in the implementation of a project in any village by contributing, cash, tools, land, materials and labor etc. this type is basically cooperation based and on limited scale.

iv. Empowering orientation. The aim of these NGOs are to help poor people an d develop a clear understanding of the social, political and economic factors which are effecting their lives, and aware them how can they solve their problem by using their resources and purpose to mobilize the people or self mobilization. In any case there is maximum involvement of the people with NGOs acting as a facilitators.

Types of NGOs by the level of operation. It has further types which are as following.

i. Community based organization (CBOs) When people start feelings that what are their needs and how can they fulfil them. These NGOs arise out of people's own initiatives. These can includes sports clubs women organizations neighbourhood organizations, religious and educational organizations. Some supported by NGOs, national and international NGOs and other independent outside help. Some are devoted to raising the consciousness of urban poor or helping them to understand their rights in gaining access to needed services while others are involvėd in providing such services.

ii. Citywide organizations. These NGOs are organized for some major or personal purpose. For example cambers of commerce and industry, coaliation of business, educational group. Some exist for other purposes and become involved in helping the poor as one of many activities, while others are created for the specific purpose of helping the poor.

iii. National NGOs. It includes organizations such as the Red cross, YMWCAs, YWCAs, professional organizations etc. Some of these have state branches and assist local NGOs.

iv. International NGOs. These range from secular agencies such as REDDA BARNA and save the children organisation, CARE, UNDP, UNICEF. Their activities vary from mainly funding local NGOs institutions and projects and implementing the projects themselves.

NGOS EMERGENCE IN PAKISTAN

In Sub-continent NGOs culture took shape in the form of Ghandi Ashram Banaras in 1927. This venture created to provide jobs for natives in the days of British Raj Diyal Sigh trust is an example of the time but these efforts could not become a social norm due to a highly centralized bureaucratic governance. As it is not a new phenomenon for Sub-continent. It emerged during the colonial period when religious, linguistic and ethic communities felt their cultural, religious and social identity threatened. Renouncing politics they concentrated on religious, cultural and social assertion.

To control these associations, the colonial authorities introduced the system of registration under the act of cooperative societies each society was required to give constitution and by laws and maintain financial accounts. The major purpose of theses organizations was to open educational, institutions, help the poor and destitute and improve the condition of women. Such welfare, charitable and educational organizations produced a breed of social workers who devote their lives to social work. They were sincere and concerned with the welfare of their community. These community based organizations also created a since of competition among each other which resulted in positive development. The great contribution of old NGOs was that they preserved cultural, social and religious values and in resistance to colonial states started movements which lead to positive struggle in the field of development.

Unfortunately, in Pakistan these NGOs cannot work properly due to political weakness' till 1958 so, under these circumstances in 70s new type of NGOs emerge which were quite different from old ones. The only thing common in old ones and new NGOs is that both came into being into response of state weaknesses. But in the absence of well-defined policy for NGOs, there is no moral considerations practiced by the people who have monopolized this vital sector. On one hand they serve as an employment exchange for kith and kin of the privileged and on other hand they are a symbol of prestige for the selected few that know the art of preparing proposals and report written in the bureaucratic lexicon. NGOs claim that there only task is to create social awareness but when people fell that NGOs are not helping them concretely, they lose interest and merely social awareness is of no use to them. Some of the clever participants turn this opportunity to their own favours by manipulating different NGOs to get funds in the name of social work. They know that projects are foreign funded and there is no commitment and sincerity behind it. Most NGOs have more or less become family business making big profit if you are a good pretender you can generate huge funds. As it is discussed above that NGOs receive funds from broad but nobody knows where and how these funds are utilized. So, people don't trust NGOs foe help as they consider them as fraud.

NGOS AND THEIR ROLE IN THE GLOBAL SOUTH

Turner and Hulme (1997) define NGOs as "associations formed from within civil society bringing together individuals who share some common purpose." Hulme (2001) characterizes them (as well as civil society) as "peopled organizations [that] are both not part of the state structures, are not primarily motivated by commercial considerations or profit maximization, are largely self-governing, and rely on voluntary contributions (of finance, labour or materials) to a significant degree." So, as Fowler (2000) observes, "for our purpose, business is not included." In support of this analysis, Edwards and Hulme (1995) expound that "most organizations referred to as NGOs thus belong, analytically, to the private sector, albeit to the service (i.e., not-for-profit) sub-sector thereof."

From a larger perspective, the prominent elements of a society are the state, the market (private sector), and the civil society – the "third sector." The state maintains public order and, to one degree or another, serves its citizens' needs. Companies in the marketplace pursue profits. And NGOs? They, like the state, seek to serve community needs, such as health, education, water, and sanitation. They may do so at least partly with government funding. For example, Clark (1991) states that "Indian NGOs are now of much significance to the country's development efforts.... The government's latest five-year plan has a one and a half billion rupee provision for funding NGOs." At the same time, though, NGOs stand apart from the state, engage in policy advocacy, and sometimes criticize government institutions and officials. As Fowler observes (1997), NGOs may provide a link between micro-level actions (providing individuals or communities with construction materials, farming equipment, or legal advice, for example) and macro-level actions (policy advocacy, lobbying, and monitoring how the state uses its powers).

Micro and macro can be viewed from a different perspective, as Turner and Hulme (1997) do: "a primary distinction can be made between organizations that are based in one country (or several countries) and [those that] seek to assist in the development of other countries. These are international NGOs (INGOs). Intermediate NGOs which operate across developing country or a region of a country [in the South can be termed] Southern NGOs. Closest to the practice of development are grassroots organizations (GROs) that operate within a limited area such as in a group of villages or part of a city."

WHY DO NGOS EXIST?

The existence of NGOs stems from both internal and external factors. Internally, the gradual retreat of the government in public service delivery has left a vacuum that NGOs try to fill. The retreat is due to governments' inability to provide high-quality public services to citizens. From after World War II to the late 1970s, the role of governments was primarily to run the public sector, oversee the economy, and treat its citizens as consumers. In development, especially in Africa, the dominant approach during this period was top-down, state-controlled, and supply-driven. As a result, citizens could not realize their potential to organize and make optimal use of their human, financial, and natural resources.

The period was characterized by malfeasance. As Tendler (1997) observes, "public officials and their workers pursue[d] their own private interests rather than those of public good." Many countries in Africa developed noticeably weak public institutions with inefficient operations, incapable of combating poverty. Hulme (2001) notes that "during the 1970s the failure of this approach rapidly to deliver economic growth and poverty reduction was increasingly acknowledged and by the early 1980s a paradigm shift was evident." Budget cuts during this period left the state increasingly "unable to cope with its basic functions of provision of infrastructure and social service".

The wave of globalization has challenged the effectiveness of the state and its bureaucratic systems, especially centralized political, administrative, economic, and fiscal systems. As the Commonwealth Secretariat (1996) argues, "the capacity of the public sector to establish the right regulatory frameworks for development, to enforce them, to develop national productive capacity, to attract capital, and to act as producer, are all in question."

Into this gap stepped NGOs, with new approaches to enhance efficiency and effectiveness in providing public services and infrastructure. At the same time, NGOs have filled a crucial role in enabling people to organize themselves and share responsibility for governance. "NGOs exist as alternatives" to a governmental, centrally led economy, in the view of Mitlin, Hickey, and Bebbington (2005). With new models of public management and many governments seemingly open to reform, the view of NGOs as alternatives is justified.

In large part, the governance reforms require the state to devolve its powers from the central government to institutions closer to the public. An important result, according to Hulme (2001), is "the return of the state for civic organizations – and particularly NGOs and GROs – with a focus on the role that they can play to improve the access that poorer and disadvantaged publics have to basic social and economic service."

Externally, the existence of NGOs has also been stimulated by increased eagerness on the part of the donor community to channel aid through them. As Hulme (2001, 137, citing Edwards, M., and Hulme, D.) argues, "the rise and rise of NGOs throughout the 1980s and 1990s was fuelled by international development agencies and aid donors who assumed that civic organizations should rapidly scale their direct service provision function". As a result, during the last two decades, both developed and developing countries have witnessed steady increases of NGOs. In the South, for example, "the number of registered NGOs in Nepal increased from 220 in 1990 to 1,210 in 1993; in Bolivia from 100 in 1980 to 530 in 1992; and in Tunisia from 1,886 in 1988 to 5,186 in 1991". NGOs have also grown more numerous in Tanzania during the last two decades. As Oda van Cranenburgh and Rolien (1995) report, "since the early 1980s and especially 1990s a considerable growth in the number of national NGOs in Tanzania has taken place [with] an increase from 137 NGOs in 1986 to 470 NGOs in the 1990s, ranging from socioeconomic development type of NGOs, professional, religious, environment, women, youth, health, education, to HIV-AIDS NGOs." The number of NGOs increased particularly during the past decade, following the approval of Tanzania's NGO policy. TANGO now includes "620 NGOs, most of which are regional and district networks that have members in the regions of 50 plus. This makes the TANGO membership by proxy to be around 1500 NGOs" (TANGO 2005).

The North, also show that NGOs have dramatically increased over the same period. "In western Europe and the USA the pattern is more complex. In America as whole the national non-profit organizations have increased from 10,299 in 1968 to almost 23,000 in 1997". The number of NGOs in the United Kingdom is even more dramatic: " Britain has a well developed voluntary sector, with a total of 200,000 registered charities in 1995" (Charities Aid Foundation, as quoted in Randel and German, 1999).

In sum, I concede to Shivji (2003, 694), who argues that "an alternative world, a better world... is possible." That is why NGOs exist. But what role and function should they adopt in the first decades of the 21st century?

The Role and Function of NGOs

In general terms, NGOs provide potent forces for social, political, and economic development. In specific terms, the literature cites a great many roles and functions of both the Southern and Northern NGOs.

Facilitating Community Programs

During the first decades of the 21st century, Southern NGOs will have to refocus their role and function. Although Lewis (2001) argues that the NGO is an implementer and "can be engaged in providing services to its clients through its own programmes," I argue, on the contrary, that the NGO should not implement its own programs, but rather should help communities achieve their own sustainable programs economic, political and social areas. As Fowler

(1997) underlines, "facilitation is a critical aspect of participation process" that Southern NGOs need to learn and practice. The term *facilitate* here refers to the process of creating space for people to act. Under this definition, how can an NGO have its own programs, unless those programs specifically seek to build local skills and capacities? As suggested by Lewis, NGOs customarily adopt a top-down and supply-driven approach to social, political, and economic development. Where NGOs directly implement their own programs, they are likely to minimize any sense of ownership on the part of the community. The top-down approach in turn locks up people's potential to act.

In that respect, community at the grassroots level may not see any need to mobilize resources and contribute toward implementing someone else's programs. They will instead simply depend on the NGOs, "the experts," to implement programs. Such NGO programs as health, water, income generating, civic education, and advocacy often unfold with minimal dialogue between the community and the NGO.

If, by contrast, community development programs facilitated by NGOs are seen as negotiated undertakings, which match local livelihood strategies and development opportunities, NGOs in the South may, as Lewis suggests, be reluctant to assume ownership of any program. By assuming ownership, Smillie (1999) argues, NGOs "are likely to make fundamental trade offs every day between the needs of their beneficiaries and the opportunities created by the emotive culture of charity." In this view, the programs, even if contracted by government or a donor agency, belong not to the NGO but to the community.

How can an NGO assume that it owns a program that is influenced and in part undertaken by the community? That is why Helmich (1999) observes, "while the practical expertise of NGO in poverty reduction is large, a gap needs to be closed between the setting of NGO goals and actual NGO practice." Optimal programs emerge when they are organized jointly, with the NGO pursuing a specific *community* goal, whether economic, social, or political. Bridging the gap requires participatory facilitative skills, which Southern NGOs too often lack. Resources that Southern NGOs mobilize from within and outside their countries may be used as seed money for the community to map out programs that will enable citizens to secure sustainable livelihoods – as Fowler (2000) puts it, "catalysing the citizen base." Southern NGOs have to also be serious about their role and function. Most of them seem to lack clear vision and mission. In fact, Shivji (2004) comments, "most NGOs do not have any grand vision of society, nor are they guided by large issues; rather, they concentrate on small, day-to-day matters. In NGOs, we seldom spend time defining our vision in relation to the overall social and economic context of our societies". Lacking vision and mission, NGOs find it difficult to articulate their facilitative role and function.

Strengthening the Capacity of Southern NGOs: The Role of Northern NGOs

In the upheavals of recent years, both developing and developed countries have witnessed three important shifts. First, governments have decentralized, particularly in developing countries, with a view toward empowering the public. Civic organizations and the market are granted considerable space. Governments in developing countries are slimming down and delegating responsibilities, so NGOs and other institutions gain increasing responsibility over political space. However, emerging institutions, both government and civic, are often weak in these countries.

Second, Southern NGOs that are keenly interested in providing an alternative approach to government are increasing and growing. These NGOs often lack the capacity to constitute a potent force in social, political, and economic development. Southern NGOs need improved capacities if they are to meet those challenging scenarios.

Third, CARE, Oxfam, Save the Children, Concern World Wide, and many other NGOs from the North are streaming to the South in the name of development. These NGOs either fund Southern NGOs or directly implement programs in the South. The Northern NGOs need to refocus. They ought to make a radical shift from implementing programs in the South to strengthening the capacity of Southern NGOs. To support the argument, Smillie (1999) provides a substantive example. The Canadian Partnership Branch "has articulated several objectives. Among them is capacity building in developing countries: to strengthen the capacity of southern organizations and institutions to make a significant and sustainable development impact among the disadvantaged communities".

Seemingly, this is a realistic and achievable objective for Northern NGOs during the first decades of the 21st century. Additionally, Northern NGOs have greater skills in fundraising than Southern NGOs. For example, Randel and German (1999) argue that "in 1996 the voluntary income of Britain's top 500 charities reached almost £2 billion (about $3 billion)." Southern NGOs can hardly dream of amassing such resources.

Northern NGOs may use their resources to strengthen the capacity of Southern NGOs on many fronts, as Duhu (2005) notes: "program support, institutional support, technical support, partnerships and coalitions." Skills in the area of strategic planning, for example, cannot be overemphasized. Sadly, Lewis (2001) argues that "in many aid-dependent contexts it is common for partnerships involving NGOs to have passive character, often because the idea of partnership is forced in some way."

There is cause for optimism. The very concept of partnership is relatively new, and Northern NGOs may not shy away from the needed changes. "Many Northern NGOs moved in recent years in broad terms from the approach in which they implemented projects themselves in developing countries... to one which most now seek to form partnership". The partnership strategy, in itself, is necessary but not sufficient.

Integrative partnership, which enables Southern NGOs to learn by doing, is the necessary approach. As James (2001) argues, "while Northern NGOs may have been instrumental in capacity building process on the agenda, ownership for the capacity building must quickly reside in the client NGO." It may empower the client to own the change process for sustainable development programs. Northern NGOs need to further search for ways to enhance effective and efficient partnership and thereby strengthen the capacity of Southern NGOs. Otherwise, as Edwards (2005) cautions, the South faces the danger of having outsiders "promot[e] certain associations over others on the basis of preconceived notions of what civil society should look like."

This chapter has argued that NGO exist for two broad reasons. Internally, they exist because the government cannot deliver high-quality public services to its citizens, leaving a space for NGOs to step in – and, ideally, to help people organize and self-develop, and to make the best use of the community's human, financial, and natural resources. Externally, NGOs exist because donors channel funds to them.

In the era of new public management, NGOs hold increasing responsibility for social, political, and economic development. To succeed, Southern NGOs must help the community implement its own vision. They must become responsible agents of change. And Northern NGOs must help them succeed.

INFORMATION ON NGOS

NGOs occupy an increasingly prominent political role, influencing policymaking in all areas of social and economic change. They are one visible manifestation of society's recognition of the inability of governments to solve major problems by legislation or command. They are also a reflection of the realisation that people can achieve more by acting together for a common cause or interest than by leaving it to the actions of governments or individuals.

Voluntary association is one of the defining characteristics of a free society. The ability of groups of citizens to organise together in pursuit of common aims and objectives is one of the features that sets democracies apart from totalitarian and authoritarian regimes.

The number of NGOs is impossible to calculate but it is safe to say it is very large. In a detailed report published in 1995, the Commonwealth Foundation estimated that in Britain alone there are more than 500,000 NGOs, and that the turnover of the 175,000 registered charities in the UK was 17 billion pounds sterling a year. It refers to one estimate that in India alone there are 100,000 NGOs, with 25,000 registered grass-roots organisations in the state of Tamil Nadu. Obviously, they vary markedly in size, resources, focus and influence. The report quoted UN Development Program estimates that the "total number of people 'touched' by NGOs in developing countries across the world is probably 250 million," although this almost certainly understates the case if account is taken of the NGO influence on public policy making.

These NGOs exercise influence and often power in our society in ways which sometimes seem disproportionate to their memberships and the weight of their arguments, and which often run counter to the wishes of the majority. Very little is known about how many of these organisations are funded, whom they represent, how they reach their decisions, and to whom they are accountable. Voting membership is usually tightly controlled, with regular contributors and donors rarely consulted about policy, strategy or tactics. Contributors always have the right to withhold further support if they disapprove of the actions of particular NGOs, but their contributions enable such organisations to claim a mandate to speak and act on behalf of a larger number of people than their actual membership.

CONSULTATION ON CAPACITY BUILDING

Nearly half of the respondents identified core funding as their highest priority need. This reflects a common concern among indigenous NGOs, whose efforts to strengthen their institutional capacity and pursue their own priorities and goals, are severely constrained by lack of resources and a consequent need to give priority to donor interests and programmes, to obtain minimal funding for their institutional survival. With very limited financial resources of their own and lack of direct access to donors, indigenous NGOs often feel their independence to be threatened. Planning and strategic development, as such, were not place among the most identified needs. However, project and programme design and implementation were recognized as important needs by a large proportion of respondents, coming second among their priority needs. At a similar lever, they emphasized priorities of networking with other NGOs, and staff development.

These three responses reflected their perceived needs in capacity building and information-sharing, to enhance their professional effectiveness. At a slightly lower level, the need for enhanced capacity in monitoring and evaluation also reflected a felt need for improved professional capacity.

At a fourth level of priority, one-third of the respondents, stressed the need for leadership development, board development and enhancement of their capacities in policy research and analysis and local resource mobilisation. These articulations recognize the importance of strengthening their organizations internally, building their capacity to understand, and hence to influence, the environment in which they work, and to contribute to their independence by mobilizing local resources to support their efforts.

Slightly fewer than one-third of the respondents prioritized the need for enhanced capacity in financial management, drawing attention to the need of many NGOs for development of simplified, easy to operate systems to facilitate their maintenance of financial accountability.

A similar priority was accorded to networking with civil society organizations (CSOs) of various types, including, but not limited to NGOs. This reflects recognition of the fact that formally organized NGOs are a small proportion of the civil society organizations in the region and that networking and collaboration with the wider category of CSOs provides NGOs with the potential to greatly enhance the impact of their work. Fewer than one-fifth of the respondents prioritized fundraising, reflecting the reality of a situation in which there are few potential local sources for significant fundraising, and in which local NGOs have little direct access to donors who usually prefer to use international NGOs as intermediaries.

Relatively low priority was accorded to organizational development, networking with northern NGOs and improving collaboration with donor agencies. These reflect their reaction to the existing situation in which they perceive that northern NGOs use their monopoly on direct access to donor funding to perpetuate dependency of indigenous NGOs, their belief that the capacity building offered by northern NGOs usually relates to their own priorities, rather than those of indigenous NGOs, and their perception that there is little that they can do to obtain direct access to the donor agencies. Fewer than one-tenth of the respondents prioritized cross-sectoral collaboration with government. This reflects the existing situation in which governments in the region tend to distrust any organized groupings beyond their direct control and regard them as potential sources of political competition. A similar level of low priority was accorded to research, documentation and perspective-building, due to the fact that most of the NGOs in the region have limited capacity in this area, and have, what seem to them, much more pressing needs. Only one of the respondents prioritized cross-sectoral collaboration with business, reflecting a situation in which business interests have thus far developed very limited social conscious and in which, they are further constrained by their vulnerability to governments which have yet to accept the role of NGOs.

Partnership Benefits

With respect to partnership issues, the most cited benefit (by 11 of 21 respondents) was that of mutual learning from exchange of experience, while 10 of 21 cited increasing programme quality and enhancing organizational and management capacities. Next was 'increasing legitimacy with other stakeholders', mentioned by 8 of 21 respondents; and 'promoting more effective advocacy' (5 of 21). At the low end of the scale: 4 of 21 respondents mentioned increasing programme scale of impact as a benefit of partnership, and 1 cited the introduction of excellent programmes.

Most Identified Issues

The most frequently identified problems related to partnership included the need to preserve NGO mission and independence (12 of 21)—reflecting often imbalanced relationships between international NGOs controlling access to funding—and local NGOs with little access to resources and the issue of establishing mutual trust and respect (10 of 21), reflecting similar problems. IN a similar context, 9 of 21 respondents cited problems of reaching agreement about cost-sharing; 6 of 21 drew attention to difficulties of reaching agreement on programme design, and programme monitoring and evaluation. Four of 21 cited problems of reaching agreement on both basic development values and development problems.

Less-cited issues included 'creating mechanisms to resolve conflict' (3 of 21); staff incompatibilities in programmes and setting priorities (3 of 21); reaching agreement on financial systems (2 of 21); and building principled alliances and bases of partnership.

Issues of partnership and the nature of partnership with external agencies and international NGOs are of particular concern to local NGOs in the region. The influx of northern NGOs to the Horn of Africa region over more than 20 years has led to one of the highest concentrations of northern NGOs in Africa. This group, however, despite its long presence in the region, has made very limited contribution to the capacity building and development of local NGOs, a type of contribution, which is vital to genuine partnership.

INFORMATION AND COMMUNICATION TECHNOLOGIES FOR IMPROVED GOVERNANCE BY BHAVYA LAL

Governance can be defined as the process through which institutions, businesses and citizen groups articulate their interests, exercise their rights and obligations and mediate their differences. Information and communication technologies (ICTs) can help to sustain this process in three ways: (i) they can support tasks that involve complex decision making, communication and decision implementation, (ii) they can automate tedious tasks done by humans, and (iii) they can support new tasks and processes that did not exist before. When ICTs are properly aligned with governance goals, they can help to create gains in both efficiency and effectiveness. There is tremendous African optimism that such gains can help address Africa's main governance challenge-how to solve grave economic and social crises with meager resources. Examples of well-thought applications around the world show that ICT can help to:

1. Reduce poverty by creating a more skilled workforce and increasing the penetration of aid and subsidies to the underserved.
2. Provide basic needs by improving the quality of healthcare, providing educational opportunities, planning for basic service delivery, and helping to improve agricultural productivity and commerce.
3. Improve public administration by facilitating informed decisionmaking, managing the burden of foreign debt, revitalizing local economies, improving policing and public safety, improving public administration and efficiency, facilitating regional, national, and sub-national coordination and communication, improving the quality of public services, and facilitating better post-conflict reconstruction and administration.
4. Enhance democratization and citizen empowerment by establishing an "open" online government, enhancing interactions between government and citizens, revitalizing civic institutions and public debate, and promoting equity and empowering minorities.

The enthusiasm for realizing the potential of ICTs in Africa is often dampened by the barriers to successful implementation. These barriers are imposed most often by lack of good infrastructure-both physical and regulatory-but also by lack of access to technology in rural or remote areas and to the poor and the underprivileged (generally women and minorities).

Lower levels of literacy, both computer-based and otherwise, and lack of content in local languages further exacerbate the difficulties.

Nonetheless, the number of governance applications is increasing, as infrastructure and literacy levels improve, and costs drop. Most governments in Africa have Web sites, and while they are still targeted toward foreign audiences, there are signs that there is tremendous progress being made in integrating ICTs in governance applications. As the number of applications increase, there are certain lessons-derived both from successes and failures-that are coming to the fore. The most important ten lessons that have been observed in governance can be summarized as follows:

- Sometimes the simplest technology can produce the biggest results.
- Using technology in governance is a trickier problem than using it in business.
- The newest or most cutting edge is not necessarily more useful or even the correct choice of technology.

Developing economies cannot afford to experiment, or to be experimental laboratories for new technologies, or for dumping excess product.

1. Generating returns from an ICT investment in the public arena requires major investment in training and support on part of its sponsors.
2. Most of the challenges do not involve the technology itself, but its alignment with existing processes and the organization's strategic goals.
3. Technology is a double-edged sword and almost always brings some challenges with it.
4. An ICT implementation that creates information haves and have-nots, either on the basis of access or computer-literacy, will eventually fail to achieve whatever goals of governance it has.
5. ICTs may not always bring about the desired benefits.
6. Finally, we must make the point that ICTs are not intrinsically valuable, or even necessary for improving governance. Given Africa-specific barriers of access and education, and general barriers of technology deployment, a framework of action incorporating these lessons learned can be useful for creating applications in governance.

The Framework Can Be Summarized as Follows:

- The first step in using ICTs as a tool to improve governance is to ignore ICTs altogether and focus on selecting and prioritizing improvement goals that are urgent or important. Once the most important goals are established, senior level policymakers must establish milestones that will indicate that the activities designed to meet these goals are on track.

- The next step is to review alternative ICT solutions that support the activities designed to achieve the goals, given constraints on financing, infrastructure, literacy and skills. Each solution must be associated with (financial and opportunity) costs-of infrastructure, training, etc.-and benefits.
- Once the ICT solution is accepted based on the planners' estimation of its merits and costs, a detailed workplan must be developed, with provisions for adequate training and capacity building. Again the key is to focus on strategic goals and user constraints.
- The final step in the process is to lay the groundwork for monitoring and evaluation (M&E) and to incorporate M&E as an ongoing integral part of the process of adapting ICTs to meet needs.

The World Bank's policy centrepiece, the Comprehensive Development Framework (CDF), also sees knowledge as central. Included in its vision for comprehensive national development strategies is a component to develop national knowledge strategies. Two other major projects, the Global Development Network (GDN) and the Global Development Gateway (GDG) are also very concerned with the development of knowledge strategies. Moreover, the development forum part of the World Bank's external website is one of the best sources of on-line sharing of development knowledge.

Across these initiatives, it is possible to discern a strong awareness of the possibilities that new information and communications technologies provide for multi-directional knowledge flows. There is also a stated commitment in the CDF, GDN and GDG programmes to capacity building in the South in the area of knowledge. However, there remain concerns within and outside the World Bank about the extent to which the new knowledge vision has been enacted or, indeed, whether there are a variety of visions that are sometimes in conflict with each other. In the on-line discussion of the GDG since July and parallel meetings, considerable concerns have been raised about the possibility of the GDG stressing particular versions of debates over others.

As the GDN develops, concerns have also been raised about the extent to which it privileges economics over other disciplines and, therefore, skews development thinking. The GDN Secretariat has sought to respond to this by entitling the December 2000 GDN conference, "Beyond Economics" but concerns remain about the ability of the GDN to develop a truly holistic view about development. The World Bank has sought to further its own learning through the development of thematic groups bringing together staff to discuss key thematic areas. Some of these groups do have excellent links to external groups and individuals but a perusal of the external websites of the education-related thematic groups gives an overwhelming sense of them being about synthesizing existing World Bank knowledge first, with external Northern knowledge sources a distant second and Southern knowledge sources hard to find at all. From the available lists of partner institutions, there appear to be few in the South.

How Does the New Development Discourse Affect Democracy?

The three case studies suggest that there is considerable complexity and tension around these two key themes of the new development cooperation discourse. In their most rigorous and radical forms, the new accounts of knowledge and partnership offer the possibility of a more democratic model of development. The knowledge account provides an opportunity to step back from universal theories and to explore good practices and real contexts, and for countries to develop their own approaches through adaptive learning. Moreover, it raises the possibility of a strengthening of Southern knowledge generation capacity and a more equitable process of multi-directional knowledge flows that include South to North and South to South flows.

The partnership account seeks to move away from the excessive conditionalities and donor direction of development in the 1980s and early 1990s. It promises national leadership of development strategies and goes beyond the state to incorporate a vision of active citizenship for development planning. It highlights the need for long-term relationships and planning and the need for a focus on processes and values. However, there are existing imbalances in knowledge and power that mean that new knowledge and partnership relationships cannot simply be willed into being. Inevitably, overall agency practices lag considerably behind the visionary edge of the most progressive agency documents. Moreover, organisational cultures and opposing mindsets mean that the new ways of thinking and doing are not dominant. Asymmetries of current knowledge production will be hard to overcome and political imperatives in the policy arena inevitably will cut across any attempts to build policy on good and diverse knowledge foundations. Development partnerships too are necessarily asymmetrical at the current juncture, given the huge disparity in resources and the often pressing need to receive aid. Moreover, the need by Northern agencies for quick disbursement will continue to undermine a focus on long-term processes and the careful development of partnerships and alliances for development.

Our case studies point to much that is positive in the statements and actions of these three agencies. In particular, they point to new understandings and new attitudes that seek to move away from donor dominance of cooperation towards local leadership. However, they also point to a continued tendency within agencies to push their own agendas more strongly than they listen to the needs and aspirations of others, and to come to a supposed dialogue with their minds already made up about what they will fund. It is important that agencies have visions and priorities but, as Karlsson (1997) and Gustafsson (1999) have written, trust and humility are also vital. Given the existing imbalance in power in development cooperation, such values from the Northern side are essential if democracy in development is to be a possibility.

A VIEW FROM SOUTHERN AFRICA BY AIDA OPOKU-MENSAH

African countries have invested very little in their information infrastructures, and in some cases have failed to create an enabling environment for the private sector. What's new with knowledge? After all, it has always been central to the development of all societies, including Africa's. What's new is the fact that information today can be moved around very quickly. And this has presented all societies with challenges and opportunities. People's access to information and the level and quality of infrastructure available to them will define-to an extent-how well societies use and adapt the increased knowledge and information.

The success of Southern African countries in strengthening their national information infrastructures will be critical in determining how well people exploit knowledge. New communications technologies hold the promise of helping to increase agricultural production, deliver better health and education services, and provide more effective and participatory governance.

That's all well and good. But just how is Africa going to get there? Money is scarce for economies crippled by external debt and desperately trying to cut back social sector spending. Infrastructure equipment is scarce and, when available, expensive. Sometimes it is plain inappropriate.

On top of all this is the lackadaisical attitude of the region's political actors, most of whom do not yet have in place any of the policies or strategies needed to tackle the many communication challenges. With the exception perhaps of South Africa, countries have invested very little in their information infrastructures, and in some cases have even failed to create an enabling environment for the private sector.

Instead, it has been typically left to international organisations and local NGOs to initiate early efforts. These include connecting Africa to the Internet with financial assistance from external donors. Many countries are also hampered by the lack of transparency and accountability in their modernisation drive. The importance of telecommunications is nothing new to the World Bank either. It has been involved in the sector for many years, although loans made in the telecommunications sector only amount to 2-4 percent of all Bank lending in Africa.

All this is not to say nothing noteworthy has happened. The process of modernising basic telecommunications systems has begun and there remains plenty of room for adopting innovative technologies to suit the region's needs. Rural "telecentres"-kiosks that offer everything from computers to telephones and email services-in South Africa and Uganda are examples of innovative projects.

South African minister for posts and telecommunications Jay Naidoo recently remarked that, "African leadership must confront a major indictment against us. Two years from the next millennium there are 700 million people on the continent and only 12 million have access to a telephone, five million in South Africa alone. "A key policy requirement is the achievement of a national communications infrastructure, essential for social and economic activity. This is important in a world where reliable and speedy communication is vital to the success of rapidly globalising trade, industry and services."

NGO's and Official Donors

In the 1950s and 1960s, Non-Governmental Organisations (NGOs) and official donors tended to pursue different development agendas. Beyond support to emergencies, they were usually disinterested in each other's activities and occasionally suspicious of the other's motives. This began to change from the early 1970s when most donors followed the earlier example of countries such as Norway and Canada in directly supporting NGO development programmes. The shift of official funding towards NGOs accelerated in the 1980s. Part of this shift is explained by the growth in emergency assistance in the period but it also reflected a growing recognition of the role of NGO programmes in meeting official aid objectives in areas such as poverty reduction, environmental conservation, health and education. This Briefing Paper focuses on the various, and changing, ways in which NGOs interact with official donors and discusses possible new directions in the relationship between NGOs and donors. The emphasis is upon development activities, rather than emergency assistance and relief. One of the most tangible indicators of growing interaction has been the change in the quantity of funds official aid agencies channel to and through NGOs for their development activities. The Organisation for Economic Cooperation and Development (OECD) indicate that the total amount of official aid going to NGOs for development in 1992/93 was US$2.2bn, while data from the World Bank put the 1992 figure at $2.5bn. Why do these figures seriously underestimate actual flows?

The financial contribution of donors to NGO development activities is commonly presented in terms of two ratios:

1. The contribution of official aid funds to total NGO income.
2. Published OECD data suggest that in aggregate about 5% of all official aid is now channeled to NGOs. Not only are these figures an underestimate, but they fail to capture wide variation among different donors in the share of official aid going to and through NGOs.

Also of importance has been the pace at which donors have increased the funds they channel to NGOs. For instance, in the ten years to 1993/94, the United Kingdom increased its official funding of NGOs by almost 400% to £68.7m, raising the share of total aid channeled to NGOs from 1.4% to 3.6%. In the same period, Australia increased its official funding of NGOs from A$20m to A$71m, raising the share of total aid going to NGOs from 1% to 6%. Similar expansion occurred in the case of Finland, Norway and Sweden from the early 1980s to the early 1990s.

Funding Arrangements

In terms of donor-NGO funding arrangements, there are variations across countries. In Australia, there are 32 different funding mechanisms through which NGOs can obtain funds from the Government. However, the dominant type of NGO activity funded by donors today remains projects and programmes put forward for funding by the NGOs themselves, and utilised for projects and programmes in particular developing countries. In the United Kingdom, this is through the Overseas Development Administration's (ODA) Joint Funding Scheme; in Sweden, through the NGO Programme; in Finland, through the NGO Support Programme. Additionally, a small proportion of donor funds are channeled through a range of international NGOs, while most bilateral donors have also provided funds to NGOs specialising in sending volunteers abroad, and to NGOs working on education and information initiatives within donor countries.

All donors have introduced criteria to determine the eligibility of potential projects put up for funding: some use sectoral specialists to review project proposals, others provide funds almost on a self-monitoring basis within general guidelines. Donors vary, too, in the share of total project costs which donors are willing to fund, from 50% or less (the UK) to 75% and upwards (Finland, Sweden). Donors have also differed in the relationship between the level of funds requested by the NGOs and the official funds available: some parliaments (Sweden) have, until very recently, repeatedly voted more funds each year than there are projects available to fund. Others (such as the UK) have to reject a high proportion of projects, because of a shortage of total funds allocated.

Different donors have also applied varying degrees of conditionality on the non-project funds they provide for NGO work. In contrast to the United Kingdom, which has stringent conditions attached to official funds used for development education and information work, other donors, such as the Scandinavians, have a more permissive approach and even provide funds for activities and campaigns critical of official aid policy.

THE ROLE OF NGOS IN DONOR PROGRAMMES

What all these particular funding schemes have in common is that they are official contributions to the NGOs' own development projects and programmes. Increasingly in recent years, however, NGOs have been co-opted to assist official aid agencies execute donors' own projects and programmes. For these types of initiative it is usual for donors to contribute all the funds required to execute these particular projects effectively on a 'sub-contract' basis. Although aggregate data on the amount of official funds channeled to these types of initiative have not been gathered, country studies conducted by the ODI indicate that, in recipient countries with a large and growing NGO presence, 5% and more of total bilateral aid funds are commonly used for these NGO sub-contracted initiatives.

There are three factors, which have influenced donors to utilise the skills and services of NGOs to help further their own agenda.

1. First, donors have been using NGOs to support their emergency and relief activities for some time, so providing funds for NGO development projects has often been viewed as a natural progression.
2. Second, poor performance of official donor programmes in reaching the poor and carrying out successful rural development projects in the late 1960s and 1970s, married with the clear popularity of NGOs for their work in the fields of education and health, and claims by NGOs that they were able to reach the poor and improve their lives, has led

donors to turn to NGOs to help them achieve a greater poverty focus in their own aid programmes.

3. Third, and relatedly, donors have seen NGOs as a means of getting around obstacles to aid impact caused by inefficient and corrupt governments, as well as a way of reaching people in those countries where they had suspended official aid programmes.

OFFICIAL DONORS AND SOUTHERN NGOS

Historically, most official funds have gone to support the work of NGOs based in donor countries, even though the bulk of the funds have been spent in developing countries. An early reason for this was that there were few viable, and effective, indigenous NGOs. Yet over the past 15 years there has been rapid growth in the number, as well as the capabilities of NGOs based in developing countries: southern NGOs.

The growth of southern NGOs has varied from country to country but, in general, effective southern NGOs emerged earliest in south and east Asia (such as Bangladesh and the Philippines) and in a number of countries in Latin America such as Chile, Brazil and Nicaragua. Only in the last ten years has there been a rapid growth in the number and importance of indigenous NGOs in Africa and their influence varies markedly across countries. The donor view which saw increasing merit in working through NGOs, together with the growing strength of southern NGOs, has led more and more donors to supplement their support of northern NGOs with direct funding of southern NGOs. Such donor-NGO initiatives became prominent in the early 1980s and have continued to expand thereafter.

In 1988, the ODA channeled £3.4m to 40 Bangladesh health-related NGOs under the Bangladesh Population and Health Consortium, and over a five year period to 1993 has provided over £5m to a large NGO in Bangladesh, the Bangladesh Rural Advancement Committee. The United States, Canada, Norway and Sweden are amongst the leading bilateral donors who have all channeled substantial funds to local NGOs, with a heavy concentration in south Asia. The EC and the UNDP have been among the leading multilateral agencies to fund southern NGO activities.

Mirroring the support given to northern NGOs, official funding of southern NGOs has taken two forms: the funding of initiatives put forward by southern NGOs, and the utilisation of the services of southern NGOs to help donors achieve their own aid objectives.

Early moves by donors to fund southern NGOs directly have often been viewed with misgivings by northern NGOs. Yet when donors have embarked on this type of initiative in consultation with their home-based NGOs, and especially when they have used the experience of northern NGO personnel on the ground to assist these direct funding initiatives, the process has often stimulated northern NGOs to assess their own comparative advantage and has been welcomed.

Donor funding of southern NGOs has received a mixed reception from recipient governments. Clear hostility from many non-democratic regimes has been part of more general opposition to any initiatives to support organisations beyond the control of the state. But even in democratic countries, governments have often resisted moves seen as diverting significant amounts of official aid to non-state controlled initiatives, especially where NGO projects have not been integrated with particular line ministry programmes.

THE RISE OF THE REVERSE AGENDA

The growth in official donor support to NGOs has not always been welcome to NGOs. Reluctance by many northern NGOs to accept large amounts of official aid funds has been based on two mutually-reinforcing ideas: that their development approach was qualitatively different from that of the official aid agencies, and that, as donors continued to apply conditions to funds channeled to NGOs, a rise in donor funding would increasingly be likely to compromise the integrity of NGO approaches to development.

Using NGOs to help achieve donors' own aid objectives only heightened these concerns, though the degree of concern has always varied across different donor countries. For example, in most Scandinavian countries, NGOs have received from the state upwards of 80% of income for their projects and most have not felt their integrity threatened as a result. Some major US NGOs, on the other hand, have refused to consider official funding. Though NGO anxiety about being over-run by the official donor agenda has persisted, the growing role and status of NGOs has fuelled a different phenomenon, increasingly referred to as the reverse agenda. This is the process whereby the approach and methods of the NGOs are now influencing the activities and perceptions of donors and official aid programmes, in some cases as a direct result of donors seeking out NGO ideas.

There are a number of ways in which this has manifested itself. For instance, some of the characteristics of the 'NGO approach' to development participatory planning, assessing a gender dimension, and concern with the environment have gradually been incorporated into mainstream donor thinking. Additionally, some donors (such as Australia, Germany, the

Netherlands and Norway) regularly seek out the views of NGOs in drawing up particular official bilateral aid programmes: Norway did this in 1993 for their programmes in Ethiopia and Nicaragua. Of particular interest has been the willingness of the World Bank (which has often attracted the hostility of NGOs) to engage in discussions with NGOs and to include some NGOs in the implementation of World Bank projects.

NGOS AND THE WORLD BANK

The World Bank has been subject to sustained criticism by some major NGOs in recent years over its handling of economic policy conditions attached to its structural adjustment loans, especially in Africa, and of its involvement in large projects, such as the Arun Dam in Nepal, which antagonise environmental groups. These twin pressures culminated in a 'Fifty Years is Enough' campaign by some environmental and developmental NGOs to coincide with the Fiftieth anniversary of the World Bank. Yet this public criticism serves to disguise a growing interaction between the World Bank and NGOs.

Until the late 1980s, NGOs played a relatively minor, and indirect, role in the work and activities of the World Bank. In the period 1973 88, NGOs were involved in only 6% of total World Bank-financed projects. Yet by 1990, NGOs were making a direct contribution to 22% of all World Bank-financed projects, and by 1994 to over 50%. Interaction with NGOs is actively encouraged not only in implementation but in the design and planning of projects. Another change has been the World Bank's growing involvement with southern NGOs in its projects. Thus whereas in the period 1973 91, 40% of NGOs involved with the World Bank were international NGOs, by 1994, indigenous NGOs represented 70% of NGOs involved in World Bank-financed projects.

At present, however, and in contrast to most bilateral donors, there are few mechanisms through which NGOs receive funds from the World Bank. Of greatest importance to NGOs are the World Bank's Social Funds, which channel resources to demand-driven sub-projects proposed by public, private or voluntary organisations. However, in July 1995 an initiative was announced which could further enhance the role of NGOs in relation to World Bank funding. The Consultative Group to Assist the Poorest of the Poor (CGAPP) is, according to the World Bank, designed to 'promote the replication and growth of NGO-managed programmes that provide financial services to the poor'. The World Bank has provided an initial capital of US$100m and other donors together are expected to contribute at least as much. The CGAPP will focus on so-called 'micro-loans' to the informal sector which were pioneered by NGOs and remain a major part of their activities.

Of particular importance is World Bank-NGO interaction in the area of policy discussion and debate. Although an NGO-World Bank Committee was formed in the early 1980s to provide a forum for policy discussion between World Bank and NGO staff, until recently there was little sign that mainstream World Bank policies were influenced by these exchanges.

In recent years, however, the World Bank has joined other donors in exposing itself to both NGO gender and participatory rural appraisal approaches; NGOs have been included in consultations on early drafts of World Bank reports and, although in a more limited way, NGO personnel have been invited to join World Bank economic mission teams. While both sides would acknowledge that there remain areas of disagreement, both would probably also acknowledge that the degree of interaction and the potential for policy change resulting from World Bank-NGO interaction is much greater today than hitherto.

The Further Expansion of a Common Agenda?

A direct effect of the growing influence of the reverse agenda has been to increase the common ground between donors and NGOs. No longer is it easy to talk of distinct differences between NGO and donor approaches to development.

One manifestation of a growing common ground has been the way that most donors have broadened their aid objectives. Thus, most donors now include poverty alleviation, concern with the environment and enhancing the status of women as major aid objectives.

Perhaps of even greater importance is that most donors now view action to enhance human rights and democratic processes as a constituent part of their development agenda. Additionally, many donors have taken up 'strengthening civil society' as a specific aid objective. This is doubly beneficial to NGOs both because NGOs are seen to constitute an important part of civil society, and because one of the core objectives of NGOs has been to work to 'empower' poor people, especially by strengthening the organisations to which poor people belong.

One concrete result of a growing overlap of objectives is that donors themselves are now increasingly willing to bring those projects and programmes, which for a long time were typically initiated by NGOs, within the umbrella of official aid. It is now not uncommon for donors to take over (and often expand) the funding of projects in the developing countries which were started and have been funded by (usually northern) NGOs.

Though these examples provide evidence of a widening cluster of initiatives where it is no longer possible to make a strong and clear distinction between donors and NGOs in terms of project approach and execution, it is important not to press the common agenda argument too far.

Thus a number of NGOs, including a high proportion of the long established and larger northern NGOs and a growing number of southern NGOs, remain wary of these recent developments and are still concerned to maintain their distance from donors. Some argue that the growing convergence of the NGO and official aid agenda could well turn out to be more a convergence of language about development than convergence in the overall approach to development. In particular many large international NGOs remain extremely critical of donors' support for economic policy reform (or structural adjustment) programmes.

The Future

The common ground between donors and NGOs can be expected to grow, especially as donors seek to make more explicit their stated objectives of enhancing democratic processes and strengthening marginal groups in civil society. However, and in spite of a likely expansion and deepening of the reverse agenda, NGOs are likely to maintain their wariness of too close and extensive an alignment with donors.

In aggregate, the direct funding of southern NGOs by donors, now emerging as a significant form of interaction, is likely to expand in the next few years, even though some donors (such as Norway) may not follow this trend.

Ironically, this expansion could well be accompanied by greater involvement of northern NGOs and northern NGO personnel, by contracting them to help administer and monitor the impact of such funds. This is in part because many officials of donor agencies often do not have the skills and expertise necessary to liaison effectively with the often small and dispersed organisations, which make up the southern NGO 'community'. It might be assumed that these trends will result in increased funding of NGOs by donors in the years ahead.

For growing numbers of northern NGOs, such an expansion would help to compensate for what appears to be falling aggregate income from private (non-official) sources (down to $5.4bn in 1994, compared with $6bn in 1992).

However a new, if very recent, phenomenon is that some donors who have provided large amounts of money to NGOs (Canada, Sweden and Finland) have announced cuts of 10% or more. In contrast, USAID intends to channel 40% of its bilateral resources through NGOs by the end of the century, up from 34% in 1994. In general, however, where donors have started to cut aid to NGOs, this has mainly been due to overall pressure on, and often absolute cuts in, the aggregate aid budget, in some cases reflecting doubts about the entire aid enterprise in the post-Cold War era.

Together these differing trends may enable donors and NGOs to cooperate even more closely than in the past. If a deeper sense of mutual interests and mutual purpose between donor agencies and NGOs does emerge, this might lead NGOs to devote less energy and fewer resources to criticising those aspects of the donor agenda they dislike, and more to building and widening the common ground they increasingly share.

NEW COMMUNICATIONS TECHNOLOGIES ARE REVOLUTIONISING ACCESS TO INFORMATION

The spread of these technologies represents other equally important changes. The spread of information technologies represents a huge growth in people-to-people communication, in effect a decentralisation of communication away from government and towards individuals. Old vertical patterns of information, symbolised by the old state monopoly broadcasting systems, are giving way to more dynamic, less predictable and much less controllable horizontal systems of communication. Political systems can no longer control the information their citizens receive, nor monitor or constrain how they communicate with each other. The capacity not just for North-South communication but for South-South communication is being transformed as people in different developing regions forge new relationships and alliances.

In short, those with access to these technologies are becoming more powerful and those who lack access are likely to become increasingly marginalised-politically and economically. Nevertheless, while the benefits of these new knowledge networks will reach many-and not just the rich-they are unlikely to reach the poor and could further skew power structures against them.

Take the Iridium system of 66 satellites which promises to deliver state of the art telecommunications from any one point on the planet to any other-a potentially ideal technology for many developing countries. When Panos interviewed people in the poverty stricken and flooded province of Bihar in

India, most were enthusiastic about the potential of such a system and about the fact that Iridium had set up a fund to provide cheap calls to some regions. But others were deeply worried. "We are fighting against the rich landlords who have grabbed thousands of acres of land, as well as against the criminal gangs which have mushroomed in this locality. This facility, if it becomes available, will only help the rich and the criminals." So says Deepak Bharti of Lok Shakti Sangathan, the "People's Power Organisation". "They are the users of cellular phones today, and will use satellite phones tomorrow. We will not. It is they who will be able to afford them and not us. And they will be used against us, to undermine us. What guarantee is there that the notorious Karia and Pappu criminal groups will not be having access to them? None whatsoever," he warns.

Improved and cheaper telecommunications could generate rural employment, could greatly enhance the integration of the rural with the national economy, improve living standards, ameliorate feelings of isolation, and potentially stem the steady migration of people from the countryside to the cities. It could also increase the gap between the poorest and the rest of society. Like anything else, these technologies in themselves are neither beneficent nor malevolent. Deploying them in ways that benefit the poor requires imaginative local policymaking, which reflects the priorities of all sectors of society. That requires informed, constructive public debate.

A VIEW FROM EASTERN AFRICA BY MELAKOU TEGEGN

The survival of humans has been based entirely on knowledge systems. Pastoralists are struggling to retain their knowledge system, which is under siege from a tide of modernisation that dismisses their system as backward or primitive. The World Bank talks about the "knowledge gap" between the North and South and considers it to be a serious handicap in the process of development. It is right, but it is only part of the knowledge story. For not only are there knowledge gaps, but there exist different knowledge-systems.

Knowledge-systems other than the dominant discourse need to be recognised not just as knowledge-systems per se, but as things that could be pivotal to the preservation of the environment and ensuring means of existence for the great many people who live on the edges of a rapidly modernising world. We appear to be moving towards a situation where a huge proportion of the information and communications industry-media (broadcast and print), film, telecommunications and advertising-is owned by a handful of global conglomerates. The power of such conglomerates is obvious, but their cultural influence is less clear cut. Many argue that they are responsible for a global dumbing down, a "McDonaldisation" of programming with Kenyan and Indonesian audiences being served by the same media diet of Jurassic Park, Oprah Winfrey and the Clinton sex fiasco. Others believe that the professional standards and dynamic competition introduced by the likes of STAR TV in Asia have reinvigorated staid national media.

And such concentration may not matter-it may be balanced by the limited, but nevertheless very real increase in people to people communication facilitated by new information technologies.

UNLOCKING THE VAST POTENTIAL OF INFORMATION AND COMMUNICATION TECHNOLOGIES (ICTS)

Information and communication technologies have always been essential for the promotion of development whether such knowledge was derived from the centuries old endowment of indigenous practices or from the latest cutting-edge technologies. Today, the technologies of the information and communication revolution are those at the cutting edge and their applications offer momentous opportunities for development. They present the developing countries with enormous opportunities and challenges, not only for accelerating their development but also in helping to bridge the economic and prosperity gaps between them and the developed countries. It also presents the developing countries with a unique opportunity to leap-frog onto a higher level of development. Some developing countries have in fact made significant strides in embracing and accessing the opportunities and applications of the new information and communication technologies.

Yet, billions still live untouched by the digital revolution. Only 5% of the world's population can claim connectivity and the greater majority of these are from the developed countries. Yet, only those countries with a significant level of development have been in a position to take advantage of the new opportunities. For the majority, the new low cost technology represents a double-edged sword.

On the one hand it holds out unprecedented opportunities for rapid development, but on the other, such technologies raise the level of competition too high for their current capabilities. Rather many developing countries are being bypassed as the tidal wave of the information revolution relentlessly sweeps across the world, thus running the increasing risk of being marginalized in the race for knowledge.

As a result, the gap between the developed and developing countries is being further aggravated by a worsening digital divide and it holds ominous consequences for employment levels, under-development and poverty. This adverse scenario could also lead to increased national and international tensions and instabilities.

We must therefore ask why and how we should redress this worsening situation. Part of the reason I believe, is that, while ICTs have vast potential for development, the reality is that to harness these forces for promoting development is a formidable and complex task that few developing countries have found a successful formula for overcoming.

First, there is the formidable expense of connectivity. These cost factors tend to inhibit the spread of information and communication technologies and undermine their universal usage. Moreover, while it may be true that certain development problems can be resolved through technological leapfrogging without having to rise through the traditional stages of development, it is also true that access to such solutions presupposes a relatively high level of development, which many developing countries simply do not have.

Unless there is affordable and equitable access and adequate connectivity for the peoples of the developing countries, the prospects of effectively participating in the knowledge economy are anything but optimistic.

4

Environment and Sustainable Development

DIMENSIONS OF SUSTAINABLE DEVELOPMENT

Sustainable development has a universal appreciation. At first sight, this is highly positive, as this could signal the entering of a holistic and responsible thinking into the world of politics and society. But as it often happens with other catch phrases that suddenly come into vogue, it fails to be translated into practice, this all the more so because sustainable development can be given several different interpretations. The earliest concept emphasized the need for economic development to be compatible with constraints set by the natural environment, one that satisfies the needs of the present generations without putting in jeopardy the satisfaction of needs of the future. More recently, it has also been stressed that economic development should be compatible with political and social institutions. So a holistic concept of sustainable development has emerged in which economic, ecological, social and political factors need to be simultaneously considered. Participation by individuals, particularly at the community level, is seen as an important means for achieving sustainable development and formulating development goals.

Of course, the concept of 'sustainable development' was not invented in the 1970s or 1980s. Certainly the names of Thomas Robert Malthus and Justus von Liebig have to appear in the early part of the pedigree of this concept. Earlier in this century, social scientists like Thorstein Veblen and economists such as A.C. Pigou drew attention to external costs of economic activities, and in 1950 K. William Kapp published a comprehensive analysis of all the important issues that since the late 1970s have staged a comeback under the name sustainable development. The term 'sustainable development' was probably coined by Barbara Ward (Lady Jackson), the founder of the International Institute for Environment and Development, who pointed out that socio-economic development and environmental protection must be linked. In 1972, a publication about the unsustainability of mainstream development, The Limits to Growth, triggered enormous fears. The United Nations Conference on the Human Environment, held at Stockholm in 1972, was the first major international discussion of environmental issues. The meeting marked a polarization between the priorities of economic growth and environmental protection. This polarization has dominated the debate between rich and poor countries and between interest groups within countries for many years and - given the results of the Kyoto Climate Conference in December 1997 - is still not fully resolved.

There are legitimate reasons for different perceptions of sustainable development and hence political priorities. Although the most significant ecological issues are of truly global importance, industrial and developing countries still have different problems. For the majority of the people affected by environmental problems in developing countries, lack of sanitation and sewage facilities, polluted drinking water, urban air pollution, shrinking water resources, and eroding topsoil are the most pressing problems. In industrial countries, where such problems have mainly been solved, the public focuses instead on issues such as depletion of the ozone layer as well as the accumulating carbon dioxide in the atmosphere and its potential impact on climatic change.

The World Conservation Strategy promoted sustainable development in 1980, as did The Global 2000 Report to the President prepared under President Jimmy Carter. The concept eventually achieved worldwide recognition and credibility with the publication of 'Our Common Future' (known as the Brundtland Report) in 1987, giving rise to an international consultation process that peaked in the 1992 U.N. Conference on Environment and Development in Rio de Janeiro. Since the early 1990s, understanding of the concept of sustainable development has been widened to include the social dimension and - through the work of Ismail Serageldin and others - has been made more dynamic, so that it involves preserving or enhancing the opportunities of future generations rather than preserving a historically given state of environmental quality or abundance of natural resources. Sustainability is to leave future generations as many opportunities as, if not more than, we have had ourselves.

Over the past few years, substantial progress has been made with regard to the greening of national accounts and hence with regard to measuring the welfare costs of resource depletion and environmental degradation. Comparing the latest set of indicators for sustainable development with the pioneering work of Irma Adelman and Cynthia Taft Morris in the late 1960s illustrates the growing degree of conceptual sophistication. The known policy instruments for environmentally sustainable development have been continuously improved to include subsidy reduction as well as targeted subsidies, environmental taxes, user fees, deposit-refund systems, tradable permits, and international offset systems. A number of elaborate case studies assess the success of different policy instruments.

Politically, the current debate on sustainable development falls into two extremes: One group continues to argue that the end is near, and that only a drastic and widespread change in human behaviour can stop the downward spiral towards self-destruction. The other camp argues that there is no reason to worry, as all trends towards a better life will continue. Experience suggests that the truth lies somewhere in between what could be called environmentalism and techneuphoria.

Most of today's available knowledge suggests that the forthcoming 10 to 30 years are crucial. We know that vital environmental assets, which are not substitutable (like the ozone layer), are being steadily destroyed, and that some of the environmental damages occurring are irreparable (e.g. extinction of species. We still lack a broader understanding of the interdependencies of complex ecosystems, but much of this knowledge will only expand as the natural environment continues to be irreversibly transformed. We would better apply the "precautionary approach", which was brought into the debate by the Rio Declaration on Environment and Development in Principle 15:

> *"Where there are threats of serious or irreversible damage, lack of full scientific certainty shall not be used as a reason for postponing cost-effective measures to prevent environmental degradation."*

GDP AND ACCOUNTING FOR SUSTAINABLE DEVELOPMENT

When we calculate GDP, we do not account for the depletion of natural resources. In normal business practice, depletion of capital assets is subtracted from the gross product to get a net product. That's done as "Net Domestic Product" (NDP) for the depreciation of capital assets, but not for environmental assets. The result is that many developing countries have a GDP which is artificially high, because they extract a lot of natural resources, count the sales price in their GDP, but never discount the value of the loss of the natural resources themselves. For example, when Indonesia sells its oil, the selling price of the oil adds into its GDP. Since a large part of Indonesia's economy is based on oil, the government can decide how much their GDP will rise by setting an extraction rate for oil. The politics of that situation is that when an election comes up, the government makes sure the GDP rises, so there's no "recession" to hinder the incumbent. The economics of that situation is that the country should not be used as a model for growth in other developing countries, since those without a large natural resource base cannot do the same tricks with their GDP.

The solution is to account for the depletion of environmental assets, by subtracting from GDP the value that the resource had before it was extracted. That is, the GDP is decreased by the value lost when the resource was extracted, since that value is no longer owned by the country. The value that a natural resource has when it is unextracted is called its "scarcity rent" (because the unextracted value reflects how scarce the resource is) or its "marginal utility cost" (or MUC). The MUC also applies to environmental amenities other than tangible natural resources, such as clean air, the recreation value of lakes, or the biodiversity provided by forests. When those amenities are lost, by polluting the air, spilling wastes into the lake, or cutting down the forest, the loss in value should similarly be subtracted from GDP. I'll discuss below how to evaluate losses to intangible resources.

If all of the MUCs are summed up, it would represent the total amount of natural resource depletion (NRD, where natural resources includes environmental amenities as well). The NRD is an adjustment to GDP which accounts for the losses of environmental assets, as capital depreciation accounts for the loss of capital assets. The definition of "sustainable development" is that the total amount of the NRD must be reinvested in reproducible capital (assuming that none of the depletion causes irreversible events, such as species extinction).

The MUC for natural resources is the price minus the marginal extraction cost, minus any other distortions which shift the price from its "shadow price," such as monopoly rents, tariff distortions, etc. The NRD for environmental amenities is based on the amount of value lost when the amenity is destroyed. There are a number of means of valuing environmental amenities (discussed below); the NRD can be simply defined as the value before the loss minus the value after the loss. More specifically, the NRD should be the environmental value without the activity causing the loss, minus the value with the activity. The distinction from the before/after definition is that we must also measure what would have happened had we not done the activity which caused the environmental degradation. For example, if we do not extract oil and burn it to create electricity, we would generate power by some other method, which would cause some other environmental degradation. This finer distinction comes into play when doing cost/benefit analyses; it is not so significant here.

To evaluate the losses in environmental amenities, we either look at the effects on marketed goods, or we use a proxy for goods which are not marketed, or we hypothesize a market where none exists. The three examples above correspond to those three levels of market evaluation, as detailed next. Of course, by using the market to evaluate environmental amenities, we are placing a human value on the environment, that is, we are assuming that the only value that matters is that which affects people. "Deep Ecologists" claim an inherent value to environmental amenities independent of their value to humans; that leaves economics and enters ethics, so I won't consider that here.

For tangible natural resources, we measure the value lost because we no longer possess them after we have sold them. The method suggested above is to sum up their scarcity rents, which indicates the value of the unextracted resources. A popular alternative method is to sum up the total asset value at the beginning of the year, subtract it from the total asset value at the end of the year, and consider the difference in value to be the loss due to extraction. The first method is difficult because one must estimate the scarcity rent in markets which are often highly distorted by monopolies (especially for oil), by export subsidies, by import tariffs, etc. Distortionary policies are especially common in developing countries, where natural resources make up a large component of the economy, so this method is most difficult where it is most important to estimate accurately.

The second method is easier to calculate, since one must only tally up the total value of the existing stocks at market value, which is straightforward for tangible resources. The weakness of that method is that if the stock changes during the year, it is interpreted as a gain in value, or a negative NRD. For example, if a new oil reserve is discovered, or if the estimate of an existing reserve size increases, then the total stock of oil may increase even if the quantity of oil extracted was far above the sustainable level. In addition, if the market price of oil rises significantly, the value of the total stock of oil at the end of the year may be higher than the value at the beginning of the year, even if the quantity is much smaller. That problem could be resolved by using economic prices (i.e., with all market distortions removed), but then this method has the same difficulties as the first method. The stock quantity problem is inherent in the method, and makes this method less valid as a measure of sustainability. To measure the value of intangible resources, we cannot use market prices, since markets for the goods themselves do not exist. From the examples above, air pollution uses an actual market price as affected by the unmarketable environmental amenity; lake recreation uses a proxy market price; and biodiversity uses an artificial market price. For each case, I'll describe the method of valuation, which includes both marketable goods and intangible goods.

For clean air, we measure the direct health benefits to the economy, by estimating the number of working days lost when the air is dirty and gets people sick. The marginal improvement in air quality causes a marginal improvement in health, and fewer employees call in sick; the losses from sick days are the cost of air pollution. Since air pollution is typically an urban problem, measuring the human health cost is usually sufficient to evaluate the entire environmental cost. For air pollution or water pollution problems which occur outside of cities, we measure the commercial loss to livestock, fisheries, agricultural production, etc. With acid rain, for example, the cost is primarily in lost revenue to fisheries (because lakes are damaged by acidification), and to farmers (because soil becomes acidified and less productive).

We could also measure the loss in value due to limited visibility, for example, from urban smog. There is no commercial loss due to decreased visibility, and no proxy goods apply, so the "contingent valuation" method (discussed below) would apply. Once again, we only measure the effect on human welfare -- we do not measure the decreased welfare of the livestock or fish, only the value loss that their decreased welfare causes to humans. Deep ecologists have their own welfare decreased by smog, and that welfare loss is captured in the same contingent valuation that measures the value that people place on visibility. A deep ecologist would presumably value clean air more highly than other people, and that value would count -- but the loss in value of clean air to animals themselves, or to the earth itself, or to God, does not count.

For a recreational lake, there is a commercial loss as well as an intangible loss. The loss of cleanliness in a lake creates a direct loss to businesses which use the lake as part of their business, such as boat rentals, fishing guides, campsites, etc. The commercial loss is measured directly, by revenues lost, just as we would measure any other direct loss. The intangible component of the loss is how much value people place on their recreation itself, that is, their utility for their enjoyment of the lake, which they do not pay for in cash terms.

For a lake's intangible recreation value, we use a proxy market, by constructing a demand curve based on what people pay to travel to the lake. The "travel cost method" uses people's willingness to pay to travel to the recreation site as an indication of their valuation. With increasing distance from the recreation site, the cost of getting there increases, for both direct expenses (automobile operating costs, tolls, etc.) and opportunity costs (value of time spent travelling). By graphing the cost of travel versus the number of people from each area, a demand curve is constructed, and then the consumer surplus loss can be estimated for marginal changes in supply of the recreation site. This method has been in use for decades and is the strongest in terms of theoretical foundations, because it uses revealed preferences (actual spending) rather than hypothetical payments.

For a forest's biodiversity value, we use a "contingent valuation" method, where people are surveyed as to how much they'd be willing to pay to preserve the forest. People are shown pictures, for example, of a thriving forest, with many species, versus a managed forest with few species. Then they are asked how much they would be willing to pay, say, in increased taxes, for the difference between the two. A demand curve is thus established again, since many people would be willing to pay a small amount, and decreasing numbers of people would be willing to pay a larger amount.

This method is theoretically weak, and expensive to implement, but the US government now accepts contingent valuation in cost/benefit analyses. Problems with contingent valuation begin with the problems of any survey -- the selection process may be biased (only those interested in biodiversity answer the survey); the answers may be strategically high (since the subjects know they don't actually have to pay); the information presented biases the results (different pictures, in the example above, would yield different results). Additional problems arise because it's unclear what is being measured -- people's preference for unspoiled forests, or people's general environmental desires, versus their value for biodiversity per se.

Nevertheless, contingent valuation is the method in use to measure all forms of value which cannot otherwise be measured. Other methods can measure commercial value (losses to business) and use value (losses to recreation). But in many cases, people do not conduct business around environmental amenities, and do not use them personally, but want the amenities to remain accessible so that they have the option to use them, or so that their descendants may enjoy them. People also value unique resources, even in the absence of the option to use them, for their existence itself -- for example, we value the Alaskan wilderness areas because we know that they are pristine, not because we ever intend to go there. Contingent valuation measures the option value, bequest value, and existence value, which no other methods do. Amorphous values such as those sound so vague that they seem to be beyond the scope of economics. But people's willingness to pay for the existence of an Alaskan park, indicates a human welfare increase. The willingness to pay for the existence of environmental amenities is demonstrated by the proliferation of environmental organizations, who earn their living from collecting donations to protect the existence of unusual species, unique natural resources, and other amenities. Amorphous though the values are, good economics must measure the increase in human welfare associated with protecting the environment.

Now let's look at how those techniques apply to GDP calculations. When oil is extracted, or timber removed from a virgin forest, or a lake sacrificed to development, only the value of the commercial goods produced is currently counted in GDP. To measure sustainable development, we should subtract from GDP all of the environmental losses which the development has caused. Extracting oil means that there is less oil in the ground -- the value that the oil would have had, if we had left it unextracted, should be discounted from GDP. Cutting virgin forests means that there is less forest, less recreational opportunity, and less biodiversity -- the loss of those environmental amenities should be subtracted from GDP as well. Development around an undeveloped lake means that the pristineness of the lake is destroyed -- the loss of people's value of the pristineness should be subtracted from the value of the development.

Environmental economists do not say that development in pristine areas should not occur, nor that non-renewable resources should not be extracted. We do say, however, that their full values should be accounted for. Failure to do so means that inefficient activities will occur, that is, activities which ignore part of their costs. Environmental degradation is a large "externality" of development. Internalizing the costs of environmental losses is necessary to properly assess which development is in the interests of people to do. Unaccounted external costs imply that the activity is inefficient, which means that people's net welfare is not maximized. Therefore, ignoring the loss of environmental amenities decreases people's overall welfare.

Accounting for the loss of environmental amenities is a key component of sustainable development. Extracting non-renewable resources at an unsustainable rate is relatively easy to measure and relatively easy to justify limiting the unsustainable activity. Destroying intangible resources is just as unsustainable -- there are a limited number of pristine lakes, and with the development of each one, the value of the remaining ones rises. Scarcity rent applies just as well to intangible amenities as it does to tangible resources. A country which spends its intangible environmental resources should discount the losses due to that spending, or they risk spending the resources unsustainably, and hence the resources will some day be gone.

The definition of sustainable development is that the loss of the limited resources is balanced by the creation of new reproducible capital. Development increases human welfare, but only if the damage caused by development does not cause more losses than gains. The means to determine if development is sustainable is to account for the environmental losses that development causes. Then environment and development will be seen as complements to one another, instead of as adver-saries.

SUSTAINABLE DEVELOPMENT IN INDIA

India's final energy demand grows faster than the development of its own national resources. Beyond, one can observe a regular growth in the intensity of polluting energy emissions of the economic activity. This is worsened by the misallocation of resources due to pricing policies, management systems, and more generally, policies that induce a lot of inefficiency and waste. To tackle the long run constraints of the present demand and supply trends, drastic changes in the management of the sector are required. Implementation of reforms began in 1991. Some options exist but a number of bold decisions still have to be taken and implemented to fulfill the energy needs of a population that has now crossed one billion inhabitants. Hence the need to contribute to the debate on sustainable development and scenarios for the twenty-first century In 1972, the then Prime Minister of India, Mrs. Indira Gandhi emphasized, at the UN Conference on Human Environment at Stockholm, that the removal of poverty is an integral part of the goal of an environmental strategy for the world. The concepts of interrelatedness, of a shared planet, of global citizenship, and of 'spaceship earth' cannot be restricted to environmental issues alone. They apply equally to the shared and inter-linked responsibilities of environmental protection and human development. History has led to vast inequalities, leaving almost three-fourths of the world's people living in less-developed countries and one-fifth below the poverty line. The long-term impact of past industrialization, exploitation and environmental damage cannot be wished away. It is only right that development in this new century be even more conscious of its long-term impact. The problems are complex and the choices difficult. Our common future can only be achieved with a better understanding of our common concerns and shared responsibilities.

Sustainable Future

Following are some perspectives and approaches towards achieving a sustainable future:

Finance

Overseas Development Assistance (ODA) is declining. The commitments made by industrialized countries at the Earth Summit in Rio a decade ago remain largely unmet. This is a cause for concern, which has been voiced by several developing countries. Industrialized countries must honour their ODA commitments. The new instruments and mechanisms, e.g., the Clean Development Mechanism, that are trying to replace ODA need to be examined closely for their implications for the developing countries. In view of the declining trend in ODA, developing countries must explore how they can finance their sustainable development efforts, such as by introducing a system of ecological taxation.

Private investment cannot replace development aid, as it will not reach sectors relevant for the poor. Such investments and other mechanisms can at best be additional to, not replacements for, development assistance. Conditions attached to financial assistance need to be rigorously scrutinized, and the assistance accepted only if the conditions are acceptable. Financial support for sustainable development programmes must not be negatively influenced by political considerations external to the objectives of the assistance.

Trade

Trade regimes, specifically WTO, are sometimes in conflict with sustainable development priorities. Imperatives of trade, and the concerns related to environment, equity and social justice however need to be dealt with independently. Environmental and social clauses, which are implicitly, or explicitly part of international agreements must not be used selectively to erect trade barriers against developing countries.

Developing countries will suffer a major trade disadvantage if the efforts to put in place globally acceptable Process and Production Methods (PPMs) are successful. Instead, existing disparities between the trade regimes and multilateral environmental agreements, such as those between Trade Related

Intellectual Property Rights (TRIPS) regime and the Convention on Biological Diversity (CBD), should be thoroughly addressed. Mechanisms to resolve such conflicts between multilateral agreements should be set up.

Technology

Developing countries need not follow the conventional path to development with regard to technologies but must use to their advantage the cutting edge technology options now available to 'leapfrog', and put the tools of modern technology to use. Mechanisms must be put in place to make available to developing countries the latest technologies at reasonable cost. Technology transfer must be informed by an understanding of its implications in the social, economic and environmental contexts of the receiving societies. Technologies must be usable by and beneficial to local people. Where possible, existing local technologies must be upgraded and adapted to make them more efficient and useful. Such local adaptations should also lead to the up gradation of local technical skills.

Local innovations and capacity building for developing and managing locally relevant and appropriate technologies must be encouraged and supported. Integrating highly sophisticated modern technology with traditional practices sometimes produces the most culturally suited and acceptable solutions, which also makes them more viable. This trend should be encouraged.

Science and Education

The paramount importance of education in effecting social change is recognized. Mainstream education must now be realigned to promote awareness, attitudes, concerns and skills that will lead to sustainable development. Basic education, which promotes functional literacy, livelihood skills, and understanding of the immediate environment and values of responsible citizenship, is a precondition for sustainable development. Such education must be available to every child as a fundamental right, without discrimination on the basis of economic class, geographical location or cultural identity.

Adequate resources and support for education for sustainable development makers of the potential of education to promote sustainability, reduce poverty, train people for sustainable livelihoods and catalyze necessary public support for sustainable development initiatives. Actions to improve their access to basic and higher education, training and capacity building must support the empowerment of women and girls. The emphasis should be on gender mainstreaming. Greater capacity needs to be built in science and technology through improved collaboration among research institutions, the private sector, NGOs and government. Collaborations and partnerships between and among scientists, government and all stakeholders, on scientific research and development and its widespread application need to be improved.

Population

With India's population crossing a billion in the year 2000, the National Population Policy announced in that year has special significance. Its change in focus from merely setting target population figures to achieving population control through greater attention to socio-economic issues such as child health and survival, illiteracy, empowerment of women, and increased participation by men in planned parenthood, gives it greater breadth and depth, thereby holding forth better promise of achieving its long-term objective of a stable population by mid-century. The official realization, that population is not merely about numbers but about the health and quality of life of people in general and women in particular, must be reinforced and sustained by an informed debate to bring key population issues into ever sharpening perspective at various levels of policy making from the national and state legislatures to local government institutions.

There is need for a better and more widespread understanding that the number of children desired by any couple depends on a large and complexly interrelated number of socio-economic and cultural factors, and that any policy action seeking to control population must seriously take all these variables into account. An important part of empowering women in matters pertaining to population is to explicitly recognize and respect their rights over their bodies and their reproductive behaviour. This recognition must permeate society in general, and religious, judicial and law-enforcement institutions in particular, through continual campaigning and dialogue.

The pursuit of population control must not be allowed to compromise human rights and basic democratic principles. Such compromises are often implicit in the disincentives aimed at controlling family size; in comments on the fertility of particular social groupings; and in the occasional demands to control in-migration to metropolitan areas. It is essential to place these matters in a balanced and rational perspective through informed public discourse supported by the wide dissemination of authentic data.

POVERTY ERADICATION AND SUSTAINABLE LIVELIHOOD

Poverty and a degraded environment are closely inter-related, especially where people depend for their livelihoods, primarily on the natural resource base of their immediate environment. Restoring natural systems and improving natural resource management practices at the grass root level are central to a strategy to eliminate poverty. The survival needs of the poor force them to continue to degrade an already degraded environment. Removal of poverty is therefore a prerequisite for the protection of the environment. Poverty magnifies the problem of hunger and malnutrition. The problem is further compounded by the inequitable access of the poor to the food that is available. It is therefore necessary to strengthen the public distribution system to overcome this inequality. Diversion of common marginal lands to 'economically useful purposes' deprives the poor of a resource base, which has traditionally met many of their sustenance needs. Market forces also lead to the elimination of crops that have traditionally been integral to the diet of the poor, thereby threatening food security and nutritional status.

While conventional economic development leads to the elimination of several traditional occupations, the process of sustainable development, guided by the need to protect and conserve the environment, leads to the creation of new jobs and of opportunities for the reorientation of traditional skills to new occupations. Women, while continuing to perform their tradi-tional domestic roles are increasingly involved in earning livelihoods. In many poor households they are often the principal or the sole breadwinners. A major thrust at the policy level is necessary to ensure equity and justice for them. Literacy and a basic education are essential for enabling the poor to access the benefits offered by development initiatives and market opportunities. Basic education is therefore a precondition for sustainable development. A sizable proportion (about 60% according to some estimates) of the population is not integrated into the market economy. Ensuring the security of their livelihoods is an imperative for sustainable development.

HEALTH AND SUSTAINABLE DEVELOPMENT

Human health in its broadest sense of physical, mental and spiritual well being is to a great extent dependent on the access of the citizen to a healthy environment. For a healthy, productive and fulfilling life every individual should have the physical and economic access to a balanced diet, safe drinking water, clean air, sanitation, environmental hygiene, primary health care and education. Access to safe drinking water and a healthy environment should be a fundamental right of every citizen. Citizen of developing countries continue to be vulnerable to a double burden of diseases. Traditional diseases such as malaria and cholera, caused by unsafe drinking water and lack of environmental hygiene, have not yet been controlled. In addition, people are now falling prey to modern diseases such as cancer and AIDS, and stress-related disorders. Many of the widespread ailments among the poor in developing countries are occupation-related, and are contacted in the course of work done to fulfill the consumption demands of the affluent, both within the country and outside. The strong relationship between health and the state of the environment in developing countries is becoming increasingly evident. This calls for greater emphasis on preventive and social medicines, and on the research in both occupational health and epidemiology. Because of the close link, there needs to be greater integration between them. Basic health and educational facilities in developing countries need to be strengthened.

The role of public health services must give preventive health care equal emphasis as curative health care. People should be empowered through education and awareness to participate in managing preventive health care related to environment sanitation and hygiene. Most developing countries are repositories of a rich resource-based health care. This is under threat, on the one hand from the modern mainstream medicine, and on the other from the degradation of the natural resource base. Traditional medicine in combination with modern medicine must be promoted while ensuring conservation of the resource base and effective protection of traditional knowledge. Developing countries should also strive to strengthen the capacity of their health care systems to deliver basic health services and to reduce environment-related health risks by sharing of health awareness and medical expertise globally.

CHANGING COUNTRIES ECONOMIC POLICIES

Environmental damage in industrialized and developing countries can be reduced with an existing package of measures that could be implemented without reducing the standard of living of the people that would be affected by them. Natural resources, such as air and water are being regarded as "free goods." Their costs are "externalized," i.e. they are paid by society in the form of damage to the ecosystems. Prices and market mechanisms must be adjusted to reflect environmental costs. They have to become an integral part of business calculations. Society must establish adequate prices for the use of goods held in common, water, atmosphere, air, and land. Governments will have to move beyond the traditional command-and-control regulatory approach and use more market-oriented solutions, which offer incentives and rewards to those who continuously innovate and improve in the area of environmental impacts. This work must be based both on the best available scientific evidence and on people's preferences and choices. When resources are priced properly, resource-intensive goods will become more expensive - and hence less attractive to consumers. Competition encourages producers to make the use of such goods cost-effective, i.e. to minimize it. To the extent that waste represents resources that have escaped from a production system, concern for costs will also encourage producers to minimize waste, especially when they pay to control it or are made liable for the damage it causes. Companies, which take their responsibility towards the environment seriously and develop better products and processes, will have a competitive edge over others. The competition inherent in open markets is the primary driving force for the creation of ecologically sound technology. Hence with an internalization of environmental, with adequate legal regulations, improved resource management (particularly with greater energy efficiency), but also with a change in behaviour patterns and different ways of defining "wealth" and "living standard", a change in our ecological (and social) course is feasible over the next 15-25 years.

The Dutch Advisory Council for Research on Nature and Environment came up with concrete data about the desired quantitative reduction in the use of eco-capacity. If a sustainable level of the use of eco-capacity should be achieved, the use of fossil fuels, metals and renewable resources as well as CO2 emissions, acid deposition and deposition of nutrients and metals must be reduced by 70 to 99% over the next 45 years. As the use of eco-capacity was and is unequally distributed among the world population, the North will have to relinquish part of its claim on the ecocapacity for the benefit of the South - this not only for reasons of equity but also from the point of view of international security. The study "Sustainable Netherlands" proposes a programme of action for the sustainable development of the Netherlands that works on similar value premises and comes up with comparable conclusions. Both studies underline the point, that time is an important factor in the sense that an early onset of action will lead to less social and economic friction.

But then, how will this be done politically? In democratic societies, politicians get elected and stay in power by voters living today. So far, it has been impossible to win elections by burdening political constituencies today with costs, shrinking options if not bans for the sake of future generations. Hence, as the Club of Rome puts it,

> *"Governments give priorities to politically useful short-term solutions and systematically neglect the longer-term perspective. As a consequence of such legacies of neglect, problems tend to become compounded and governments fall into a rhythm of crisis government".*

There is a tendency to treat sustainable development as merely a variation of the prevailing approaches to development and to see sustainability as a goal that can be attained through making adjustments to the standard development models. Enlightened heroism on the side of professional politicians to do the right thing-and to run a high risk of being ejected from power - is not a widespread quality. Thus, also here, "development from below" is necessary.

ENVIRONMENTAL ECONOMICS

Environmental economics is a subfield of economics concerned with environmental issues. Quoting from the National Bureau of Economic Research Environmental Economics program: "Environmental Economics undertakes theoretical or empirical studies of the economic effects of national or local environmental policies around the world. Particular issues include the costs and benefits of alternative environmental policies to deal with air pollution, water quality, toxic substances, solid waste, and global warming."

Topics and Concepts

Central to environmental economics is the concept of market failure. Market failure means that markets fail to allocate resources efficiently. As stated by Hanley, Shogren, and White (2007) in their textbook *Environmental Economics*: "A market failure occurs when the market does not allocate scarce resources to generate the greatest social welfare. A wedge exists between what a private person does given market prices and what society might want him or her to do to protect the environment. Such a wedge implies wastefulness or economic inefficiency; resources can be reallocated to make at least one person better off without making anyone else worse off." Common forms of market failure include externalities, non excludability and non rivalry. Externality: the basic idea is that an externality exists when a person makes a choice that affects other people that are not accounted for in the market price. For instance, a firm emitting pollution will typically not take into account the costs that its pollution imposes on others. As a result, pollution in excess of the 'socially efficient' level may occur. A classic definition is provided by Kenneth Arrow (1969), who defines an externality as "a situation in which a private economy lacks sufficient incentives to create a potential market in some good, and the nonexistence of this market results in the loss of efficiency." In economic terminology, externalities are examples of market failures, in which the unfettered market does not lead to an efficient outcome.

Common property and non-exclusion: When it is too costly to exclude people from accessing a rivalrous environmental resource, market allocation is likely to be inefficient. The challenges related with common property and non-exclusion have long been recognized. Hardin's (1968) concept of the tragedy of the commons popularized the challenges involved in non-exclusion and common property. "commons" refers to the environmental asset itself, "common property resource" or "common pool resource" refers to a property right regime that allows for some collective body to devise schemes to exclude others, thereby allowing the capture of future benefit streams; and "open-access" implies no ownership in the sense that property everyone owns nobody owns. The basic problem is that if people ignore the scarcity value of the commons, they can end up expending too much effort, over harvesting a resource (e.g., a fishery). Hardin theorizes that in the absence of restrictions, users of an open-access resource will use it more than if they had to pay for it and had exclusive rights, leading to environmental degradation.

Public goods and non-rivalry: Public goods are another type of market failure, in which the market price does not capture the social benefits of its provision. For example, protection from the risks of climate change is a public good since its provision is both non-rival and non-excludable. Non-rival means climate protection provided to one country does not reduce the level of protection to another country; non-excludable means it is too costly to exclude any one from receiving climate protection. A country's incentive to invest in carbon abatement is reduced because it can "free ride" off the efforts of other countries. Over a century ago, Swedish economist Knut Wicksell (1896) first discussed how public goods can be under-provided by the market because people might conceal their preferences for the good, but still enjoy the benefits without paying for them.

Valuation

Assessing the economic value of the environment is a major topic within the field. Use and indirect use are tangible benefits accruing from natural resources or ecosystem services. Non-use values include existence, option, and bequest values. For example, some people may value the existence of a diverse set of species, regardless of the effect of the loss of a species on ecosystem services. The existence of these species may have an option value, as there may be possibility of using it for some human purpose (certain plants may be researched for drugs). Individuals may value the ability to leave a pristine environment to their children. Use and indirect use values can often be inferred from revealed behaviour, such as the cost of taking recreational trips or using hedonic methods in which values are estimated based on observed prices. Non-use values are usually estimated using stated preference methods such as contingent valuation or choice modelling.

Contingent valuation typically takes the form of surveys in which people are asked how much they would pay to observe and recreate in the environment (willingness to pay) or their willingness to accept (WTA) compensation for the destruction of the environmental good. Hedonic pricing examines the effect the environment has on economic decisions through housing prices, travelling expenses, and payments to visit parks.

Solutions

Solutions advocated to correct such externalities include:

- *Environmental regulations*. Under this plan the economic impact has to be estimated by the regulator. Usually this is done using cost-benefit analysis. There is a growing realization that regulations (also known as "command and control" instruments) are not so distinct from economic instruments as is commonly asserted by proponents of environmental economics. E.g.1 regulations are enforced by fines, which operate as a form of tax if pollution rises above the threshold prescribed. E.g.2 pollution must be monitored and laws enforced, whether under a pollution tax regime or a regulatory regime. The main difference an environmental economist would argue exists between the two methods, however, is the total cost of the regulation. "Command and control" regulation often applies uniform emissions limits on polluters, even though each firm has different costs for emissions reductions. Some firms, in this system, can abate inexpensively, while others can only abate at high cost. Because of this, the total abatement has some expensive and some inexpensive efforts to abate. Environmental economic regulations find the cheapest emission abatement efforts first, then the more expensive methods second. E.g. as said earlier, trading, in the quota system, means a firm only abates if doing so would cost less than paying someone else to make the same reduction. This leads to a lower cost for the total abatement effort as a whole.
- *Quotas on pollution*. Often it is advocated that pollution reductions should be achieved by way of tradeable emissions permits, which if freely traded may ensure that reductions in pollution are achieved at least cost. In theory, if such tradeable quotas are allowed, then a firm would reduce its own pollution load only if doing so would cost less than paying someone else to make the same reduction. In practice,

tradeable permits approaches have had some success, such as the U.S.'s sulphur dioxide trading program or the EU Emissions Trading Scheme, though interest in its application is spreading to other environmental problems.

- *Taxes and tariffs on pollution/Removal of "dirty subsidies"*. Increasing the costs of polluting will discourage polluting, and will provide a "dynamic incentive", that is, the disincentive continues to operate even as pollution levels fall. A pollution tax that reduces pollution to the socially "optimal" level would be set at such a level that pollution occurs only if the benefits to society (for example, in form of greater production) exceeds the costs. Some advocate a major shift from taxation from income and sales taxes to tax on pollution-the so-called "green tax shift".
- *Better defined property rights*. The Coase Theorem states that assigning property rights will lead to an optimal solution, regardless of who receives them, if transaction costs are trivial and the number of parties negotiating is limited. For example, if people living near a factory had a right to clean air and water, or the factory had the right to pollute, then either the factory could pay those affected by the pollution or the people could pay the factory not to pollute. Or, citizens could take action themselves as they would if other property rights were violated. The US River Keepers Law of the 1880s was an early example, giving citizens downstream the right to end pollution upstream themselves if government itself did not act (an early example of bioregional democracy). Many markets for "pollution rights" have been created in the late twentieth century—see emissions trading. The assertion that defining property rights is a solution is controversial within the field of environmental economics and environmental law and policy more broadly; in Anglo-American and many other legal systems, one has the right to carry out any action unless the law expressly proscribes it. Thus property rights are already assigned (the factory that is polluting has a right to pollute).

Relationship to other Fields

Environmental economics is related to ecological economics but there are differences. Most environmental economists have been trained as economists. They apply the tools of economics to address environmental problems, many of which are related to so-called market failures—circumstances wherein the "invisible hand" of economics is unreliable. Most ecological economists have been trained as ecologists, but have expanded the scope of their work to consider the impacts of humans and their economic activity on ecological systems and services, and vice-versa. This field takes as its premise that economics is a strict subfield of ecology. Ecological economics is sometimes described as taking a more pluralistic approach to environmental problems and focuses more explicitly on long-term environmental sustainability and issues of scale.

Environmental economics is viewed as more pragmatic in a price system; ecological economics as more idealistic in its attempts not use money as a primary arbiter of decisions. These two groups of specialists sometimes have conflicting views which may be traced to the different philosophical underpinnings.

Another context in which externalities apply is when globalization permits one player in a market who is unconcerned with biodiversity to undercut prices of another who is-creating a "race to the bottom" in regulations and conservation. This in turn may cause loss of natural capital with consequent erosion, water purity problems, diseases, desertification, and other outcomes which are not efficient in an economic sense. This concern is related to the subfield of sustainable development and its political relation, the anti-globalization movement. Environmental economics was once distinct from resource economics. Natural resource economics as a subfield began when the main concern of researchers was the optimal commercial exploitation of natural resource stocks.

But resource managers and policy-makers eventually began to pay attention to the broader importance of natural resources (e.g. values of fish and trees beyond just their commercial exploitation;, externalities associated with mining). It is now difficult to distinguish "environmental" and "natural resource" economics as separate fields as the two became associated with sustainability. Many of the more radical green economists split off to work on an alternate political economy.

Environmental economics was a major influence for the theories of natural capitalism and environmental finance, which could be said to be two sub-branches of environmental economics concerned with resource conservation in production, and the value of biodiversity to humans, respectively. The theory of natural capitalism (Hawken, Lovins, Lovins) goes further than traditional environmental economics by envisioning a world where natural services are considered on par with physical capital.

The more radical Green economists reject neoclassical economics in favour of a new political economy beyond capitalism or communism that gives a greater emphasis to the interaction of the human economy and the natural environment, acknowledging that "economy is three-fifths of ecology"-Mike Nickerson.

These more radical approaches would imply changes to money supply and likely also a bioregional democracy so that political, economic, and ecological "environmental limits" were all aligned, and not subject to the arbitrage normally possible under capitalism.

Professional Bodies

The main academic and professional organizations for the discipline of Environmental Economics are the Association of Environmental and Resource Economists (AERE) and the European Association for Environmental and Resource Economics (EAERE). The main academic and professional organization for the discipline of Ecological Economics is the International Society for Ecological Economics (ISEE).

GENERAL PRINCIPLES OF SUSTAINABLE DEVELOPMENT

In addition to the five basic conditions necessary for ensuring sustainable development, five major principles underlie its implementation. These principles, to an even greater extent than the aforementioned conditions, are vital to the definition of sustainable development.

Environmental and Economic Integration

The environment and the economy are obviously very closely related. This link is more than a mere principle; it is a necessity for sustainable development. Various economic tools and policies may promote sustainable development, or at least lead to a more environmentally conscious use of resources. These tools or policies, such as the polluter-payer or consumer-payer approach, may be applied equally to producers, consumers and taxpayers and to enable the market to determine the correct overall cost of using resources. In many instances, however, for the actual value of natural resources to be taken into account, producers and economic agents need to change their attitudes. As a result, tax incentives or other economic tools may be necessary to promote this coming together of the environment and the economy. The integration of the environment and the economy is as advantageous for poorer countries as for rich ones because, if production models adhere to economic and environmental rules, there may be a better balance of comparative production advantages. The result could be a softening of world trade rules whereby poorer countries would be enabled to lay claim to greater economic development.

Certain traditional economic indicators may also assist in assessing the degree to which the economy and the environment are integrated. Particular examples are the gross domestic product and per capita income; global indicators that reflect social aspects (such as the Human Development Index, which includes longevity, education and income); and strictly environmental indicators, such as water quality and land use.

Conservation of Natural Resources

Achieving sustainable development presupposes that we can preserve biological diversity, maintain ecological processes and life support systems and use the world's species and ecosystems in a sustainable manner. Development based on the preservation of natural resources calls for energetic measures that will make it possible to protect the structure, functions and diversity of the natural systems on which life depends. These measures must focus on species and ecosystems as well as on their genetic heritage. Consequently, the limits, on and the capacity for renewal of, natural resources such as soil, wild and domesticated species, forests, pasture and farm land, fresh water and marine ecosystems, must not be compromised. As well, the life of non-renewable resources should be extended by developing and using more effective and cleaner technologies and by encouraging re-use and recycling.

First of all must come changes in the behaviour of individuals and communities and in their attitude to the environment, along with the provision of genuine means for managing it better. New approaches at the state level must then integrate development and conservation of resources on the basis of sufficient information and knowledge and through appropriate legal and institutional instruments. Effort at the international level must be on promotion of the development, and adoption and implementation of conventions and protocols on the environment and natural resources.

Precaution, Prevention and Evaluation

Precaution, prevention and evaluation are the starting points for genuine sustainable development; they must form an integral part of the planning and implementation of every development project. Planners and decision-makers must make it a routine to foresee and provide for the environmental consequences of their projects.

Current environmental protection measures are precautionary; however, in many cases, they are merely a band-aid solution that is not always compatible with the concept of sustainable development, particularly from a long-term perspective. However, the concepts of precaution, prevention and evaluation are difficult to instill because they are often removed from the day-to-day reality and have benefits that will be felt only in the more or less distant future. Forewarned is forearmed, foresight is knowledge and evaluation enables planning: it is imperative that countries and societies adopt these three watchwords so that present development can be transformed into sustainable development.

Cooperation, Partnership and Participation

Achieving sustainable development has become a collective responsibility that must be fulfilled through action at all levels of human activity. Consultation and cooperation in all decision-making are essential to the sustainable management of terrestrial, aquatic and marine ecosystems. It is incumbent upon all states and all nations to cooperate in good faith and in a spirit of partnership in implementing effective strategies to protect, preserve and restore the environment. All must take an active part and do their fair share in accordance with their capabilities and the means at their disposal.

All governments must accept their responsibilities by introducing economic growth policies and programs compatible with the protection of their own environment and that of others. They must ensure the protection of ecosystems of particular importance for agriculture and the way of life of the populations that depend on it. Furthermore, they must facilitate the participation of non-governmental organizations and decentralized or local communities to ensure they can play a greater role in all development-and environment-related activities.

In addition, states must join forces to strengthen international law by adhering to existing environmental conservation and management conventions and protocols and by passing the necessary statutes for their implementation. They must also promote and develop new agreements and instruments considered necessary to achieving sustainable development.

Cooperation and partnership also presuppose that the richest countries introduce financial and technical assistance measures that will enable the poorer countries to integrate environmental issues more easily into their development programs. The creation of specific environmental protection and restoration funds is certainly worth considering.

The preservation of biological diversity clearly illustrates how interdependent are the "North and South blocs" in the necessary establishment of new partnerships. The main "centres or sources of biological diversity" are situated more particularly in the countries of the South, whereas the major "technological or biotechnological centres" are mainly in the countries of the North. In other words, the countries of the South as well as those of the North must be party to all discussions, solutions and conventions necessary to the achievement of sustainable development. They must all ensure that the measures chosen are suited to the situation of each. The more developed countries will no doubt have to make the necessary efforts to bring about a higher degree of development in the poorer countries and, in particular, the latters' improved access to the most suitable technologies.

Education, Training and Awareness

Safeguarding the environment and achieving sustainable development depend not only on technical and economic matters, but also on changes in ideas, attitudes and behaviour. The direct participation of individuals and communities is essential. All must become fully aware of their environment, know its demands and limits and alter their habits and behaviour accordingly. To this end, countries must develop strategies to better educate, inform and sensitize their populations on environmental matters and sustainable development. For example, ecological and environmental concerns can be integrated into school programs; the awareness of the general public can be raised through extensive information campaigns, particularly through the media; "green" projects can be encouraged in local communities, and training programs can be developed to promote more informed resource management and the use of clean technologies.

SUGGESTIONS FOR SOCIAL DEVELOPMENT

There are several important issues which require detailed deliberation in social development.

The Issues are:-

1. Considering the fast growth of social development sector in 21st century there is need for Government of India to take a proactive role for multisectoral coordination and convergence of various role players involved in social development for which several Ministries of Govt. of India are required to be sensitized.
2. Need for greater transparency and Accountability in the sector. The world's leading human rights, environmental and social development international organizations such as Action Aid International, AMNESTY International, and Green Peace International, Oxfam International,

Save the Children International and World YWCA have today publicly endorsed the first global accountability charter for the non-profit sector to act as responsible players for social development.

International NGOs play an increasingly influential role. Global public opinion surveys show higher trust in NGOs than in government and business. In addition to an internal desire to be transparent and accountable, the accountability charter also seeks to demonstrate that NGOs deeply value public trust

In an unprecedented step, international civil society organizations have come together to demonstrate their commitment to transparency and accountability. This initiative builds on the individual, national and sectoral initiatives taken by international NGOs to set standards of accountability and codes of conduct.

3. Need for greater respect, appreciation and support to the social workers. The reason for this they are the key actors for the overall development. Every part of social development sector depends upon their ability and responsibility. Efficient and effective social workers will be made by the concerned NGOs. NPOs, NGOs and other developmental organizations and foundations are needed to respect and appreciate the social workers.

   The lack of social workers and the continuous high turnover of social workers weaken the quality and availability of the service given to clients especially for the poor. The lack of competent personnel may lead to the loss of the basic social rights intended in the constitution. Reasons for the lack of social workers are low pay, lack of leadership and heavy workload.

   In order to improve the availability of social workers and their quality work it is suggested that social workers should have the opportunity to participate in continuing education and supervision of work.

4. Need for further promotive policies of activities self groups, micro credit & market avenues for the productivity self help groups. NGOs can promote the policies for self help groups and micro finance. There is strong relation between the self help groups and micro credit system at present scenario.

   In India self help groups are extensively working as primary tools towards poverty alleviation and empowerment. National and state government initiatives, as well as NGOs efforts, have used SHGs to implement poverty alleviation programmes in Andhra Pradesh since 1979. Micro credit is emerged as strong weapon to eradicate poverty through self sustainable with the help of NGOs, micro finance institutions and banks. At present scenario of social development with relation to

5. Need for clear distinction and clarity of Micro Credit & Micro Finance. Mainly SHGs are the prime clients for the Micro and Micro Finance. Banks or financial institutions need to clarify the difference between these two things, because banks are main sources for the micro credit or finance.

   Micro credit and micro fiancé, often used synonymously, is very popular terms in recent developmental activities. This is creating huge confusion and misunderstanding in developmental activities. Professor Muhammad Yunus mentioned this problem with some sarcasm in his address to the International Seminar on Attacking Poverty with Micro Credit, held in Dhaka on 8 and 9 January 2003: The word micro credit did not exist before the seventies. Now it has become a catchword developmental practitioner. In the process, the word now means everything to everybody.

   Micro credit caters commercial needs of poor for enabling them to raise their income levels and improve standard of living. Micro credit means more emphasis on loans while micro fiancé also includes support services where you open up channels for thrift, market assistance, technical assistance, capacity building, insurance, social and cultural programmes. So where there is micro finance is credit plus, there only micro credit is credit.

   Micro credit financial requirements are generally not meant for economic development activities, but for consumptive needs like it education of a child, medicinal requirements etc. Here quantum are quite low, needs are very emergent, and there is hardly any difference between the consumptive purpose and productive purpose.

6. Need for coordinating Agencies & mechanism both at Govt. of India and at various State Govt. levels. Need for coordinating Agencies and mechanism both at Government of India and at various State Government level. Coordinating agencies and mechanism helps in brining the all NGOs together and facilitate for better work in social development with the help of government machineries.

   More over, these instruments helps to form networks at all level including international, national, state and local level. This body can also coordinate NGO movements in each country.

7. Need for enabling policies both by centre and states. The social problems of contemporary India are the result of a complex nexus

between the factors of exclusion and inclusion rooted in history, values, and cultural ethos. Many of these problems could not addressed by the development strategy launched since independence. Recent policies of globalization have further undermined the role larger societal norms as well as the state apparatus that could counter exclusionary forces. The agenda of social development has remained unfinished, keeping social tensions simmering.

During the 7th five-year plan, polices were helped to achieved the targeted social development goals, in terms of establishment of social infrastructure, especially in rural areas. The 8th five year plan identified "human development" as its main focus, with health and population control listed as two of six priority objectives. It was emphasized that health facilities must reach the entire population by the end of the 8th plan. The plans also identified people initiative and participation as a key element. With the enactment of the 73rd Constitutional Amendment Act (1992), Panchayati Raj Institutions (PRIs) were revitalized and a process of democratic decentralization ushered in, with similar provisions made for urban local bodies, municipalities and nagar palikas.

Today, however, in the policy debate, ongoing orthodox economic liberalism is giving way to concerns regarding social consequences of globalization, as it affects the poorest and the marginalized sections of the population. Thus, a number of highly important and far-reaching social policy measures have been brought on to the development agenda, in the form of the right to information act, rural employment guarantee act, the rural health mission among others.

8. Need for capacity building of Govt. Officials, NGOs and gross root level activists and stake holders of social development. Several international and national conferences identify an effective leadership role of the NGOs as social development channels.

Capacity building can be defined as "development, fostering and support of infrastructure, resources and relationships for NGOs and related systems and services, at Member States, organizational, inter-organizational, and regional and systems levels, contributing to the peaceful, socially distributed and sustainable development of our societies." Capacity-building programmes broaden and strengthen the professional expertise and accelerate progress in the organization activities weather it government or non governmental organization. Capacity building's main goals are to increase the individual capacity of present and future developmental professionals and leaders and to support the development of institutions and programmes all over the world especially in the social development sector.

9. Need for process documentation, Action research, Monitoring & Evaluation and social audit.

   The word "documentation" includes both records and documents. Records are recorded information, regardless of the medium or characteristics, made or received by an organization that is useful in the operation of the organization. Documents explain what an organization plans to do and how it will be accomplished as well as instruct employees how to perform tasks. In this regard professional organizations need to maintain quality and effective documentation along with the action research, monitoring and evaluation and social audit.

   Action research is very important component for the effective functioning of the organization. Action research finds out the achievement and failures of the organizations activities. With the help of action research findings, NGOs and other developmental organizations can reform or modify their on going developmental projects, if there are any insufficient methods or policies, social audit is also one of the important elements for the social development. NGOs are the prime players to enhance the awareness about the social audit within the NGOs and other governmental organizations.

10. Need for further promotion of participatory approaches of the development.

Participatory approach of the development facilitates the local communities play a central role in the planning, implementation and funding of activities within participatory developmental programmes. The exact composition of any given programme should be determined in conjunction with them. It is important to ensure that programme activities:

1. do not provoke conflict between resource users (where conflict is unavoidable, conflict resolution mechanisms should be specified early on);
2. do not further isolate marginal households (that may not be able to participate in activities which demand a labour or financial contribution);
3. do not undermine viable indigenous soil and water conservation techniques;
4. are informed by an understanding of existing management practices (e.g. they do not immediately promote group activity if there is no history of communal working);

5. are feasible given current capacity within the community and external organizations; and
6. take into account underlying climatic, hydrological, soil and land use characteristics.

Participatory approaches are more important to succeed the developmental programmes. Participatory approach also enables the social capital for social development.

11. Need for sharing of knowledge, innovative approaches, and people centred approaches. Here, sharing of knowledge is very important component in every aspect of development field. This concept of sharing of knowledge leads to innovative methods in development sector with development organizations. People centred approach, at present, is very much needed in the field of social development. These above components considered as critical elements for speedy social development. NGOs and other developmental organizations need to develop an innovative approach for social development with the help of sharing of knowledge concept. To get access of sharing of knowledge and to develop innovative approach, NGOs and other developmental organizations need to get extensive trainings on various issues such as Organizational Development, Capacity Building of Organizations and other related to social development. Moreover, NGOs are considered by international developmental organizations and banking as the prime actors for social development.
12. Need for savings, economy of resources and no cost, low cost and cost effective approaches. There is interrelation between the above concepts such as savings, economy of resources and no cost, low cost and cost effective approaches in social development. Developing the society with all minimum needs required the tremendous savings in all sections of the society. Especially in rural areas, women self help groups savings are reached at maximum level. This saving came from only marginalized sections of the society. In this scenario, NGOs and other developmental organization's efforts utilized cent percent with no cost and low cost effective approaches. Thus, for overall social development need the above concepts. Here, NGOs and other voluntary or developmental organizations need to enhance the awareness on above issues for better social development.
13. Need for exchange visits, study tours, & National and International exposure. Exchange visits and study tours at national and international exposure will broaden and strengthen the organizational capacity in various aspects such as management skills, developmental methods

and interaction with developmental professionals in the field of national and international and so on. Moreover, tours and visits also give knowledge in the different areas of social development from different nations especially from developed countries. The developed nation's technological knowledge will help in faster the social development in developing countries.

14. Need for planning by NGOs working in the sector. As earlier stated by international developmental organizations, NGOs are the critical players in social development. Because, NGOs works in the every aspect of society and they works from grass roots level to international level. So, NGOs can draw the plans for social development effectively.
15. Need for introduction of managerial inputs to the NGOs working in the sector. For the effective programme implementation and policy-making, NGOs need to acquire the knowledge of effective managerial inputs for the NGOs in the different areas of social development. In the present scenario of social development, every programme of poverty alleviation in rural as well as urban areas and infrastructure development such as roads, drinking water, schools and health centres etc., are implemented and monitored by NGOs. Thus, for effective implementation or benefits reaching to the poor are depends up on NGOs effectiveness.
16. Need for sustainable development approaches. Need for collaboration among peoples institutions, NGOs and Government Institutions. The international community has recognized the vital importance of cooperation between government agencies and non-governmental organizations (NGOs) in addressing the social and human development related issues. The importance of government-NGO cooperation was stressed in the recent global and regional conferences concerning the social development. To respond to these mandates, international developmental organizations formulated the process of government-NGO collaboration for social development especially in the area of poverty alleviation programmes. Moreover, World Bank also is stressed on the Government and NGO collaboration for rapid social development in recent international conferences.
17. Need for Multisectoral coordination and convergence of various role players in the sector. Multi sector coordination and convergence is the effective method for rapid social development. Governments across the world are grappling with appropriate policies to optimize the benefits associated with convergence through multi sectoral coordination. Convergence has emerged as a global phenomenon as

a result of digitization which has allowed traditionally distinct services to be offered across interchangeable platforms. These technological trends have been accelerated by the liberalization of markets allowing for the social development.

18. Need for integrated Micro Planning at village level and Holistic development. Integrated Micro planning is the most important concept for the developmental and governmental organizations. In this aspect, every resource of the village utilized for the overall development of the village through the micro planning. In this process, natural resources and human resources utilized for overall development with the help of NGOs and government organizations.
19. Need for creation of the Model Villages. Model villages especially created for the utilization of information and communication technology. In the previous years, ICT was utilized for the development of corporate industries and other large size multi national corporations. Now, in the globalized era, ICT is reaching to the every corner of the globe including villages. If ICT need to utilize properly in village development, villages need to be developed as model villages so as to accept the access of developed technologies. NGOs and other developmental organizations need to be developed the facilities in rural areas with coordination with the government organizations.
20. Need for non exploitation, lack of corruption, and lack of hypocritic approaches by a few black sheep in social development. As late Prime Minister of India, Rajeev Gandhi noticed that exploitation and corruption are the major hurdles for the path of India's development. This is indicating that the high intention and impact in the social development. NGOs and other developmental organizations and charities and every social group are need to work for the non exploitation, lack of corruption, and lack of hypocritic approaches for the better society and social development.

The inter-relationship between the various activities has to be emphasized and the necessary coordination assured both in the Central Government; and in the States. One aspect of this coordination would be to secure that legislation relating to social problems follows broadly similar principles.

In cases where grants-in-aid are given by a State authority to a private agency, it is desirable to lay down general directions for improving the content of the programmes and their administration. A measure of supervision and inspection should also be provided in order to maintain standards of efficiency. A major responsibility for organizing activities in different fields of social welfare, like the welfare of women and children, social education, community organization, etc., falls naturally on private voluntary agencies. These private agencies have for long been working in their own humble way and without adequate State aid for the achievement of their objectives with their own leadership, organization and resources.

Any plan for the social and economic regeneration of the country should take into account the service rendered by these private agencies and the State should give them the maximum cooperation in strengthening their efforts.

Public cooperation, through these voluntary social service organizations, is capable of yielding valuable results in canalizing private effort for the promotion of social welfare.

One of the most important tasks of the State is to conduct a survey of the nature, quality and extent of service rendered by voluntary agencies in different parts of the country, to assess the extent of financial and other aid that they are in need of in order to develop their programmes of work, and to coordinate their activities.

CONSENSUS FORMATION OF BUSINESS AND INDUSTRY

Throughout this process of consensus formation, business and industry exerted a structuring influence. They succeeded in making their view hegemonic, and ended up being considered post-Rio as a major social actor providing solutions to the global ecological crisis. As influential economic agents, transnational corporations (TNCs) have activities that directly impact on the situation of the environment. TNCs have been a constant target of NGOs, which point out their preponderant role in environmental degradation. Several public campaigns and boycotts have been organized to draw the public's attention on the issue and force TNCs to comply with legislation, adopt higher environmental standards or change production processes.

On the issue of tropical deforestation for example, NGOs have pointed out that corporations such as British Petroleum, Shell or Mitsubishi bear a large responsibility for forest devastation worldwide. Already in 1989, The Sunday Times directly accused British Petroleum and Shell of contributing to the depletion of the Amazonian rainforest in Brazil.

More recently, the Rainforest Action Network (RAN) accused Mitsubishi, together with its subsidiary Meiwa, of being "the greatest corporate threat to the world's tropical, temperate and boreal forests." RAN accuses Mitsubishi of illegal logging, transfer pricing, tax evasion, violations of pollution standards, anti-trust activity, violation of native land claims, and employment of illegal aliens. Yet despite evidence of the role of corporations in environmental degradation, the issue was scarcely discussed and questioned during the UNCED process. There is, it is true, a chapter in Agenda 21 dedicated to the role of business and industry. Yet the document does not in any way blame business for its major contribution to the ecological crisis. Agenda 21 contents itself with providing guidelines to firms in order to help them improve their environmental records.

But this is not to say that business and industry were absent or uninterested in the negotiation. On the contrary, large corporations were very active in the UNCED process, and even before it. As early as 1984 a World Industry Conference on Environmental Management (WICEM I) had been organized in France to recommend actions to include environmental concerns in industry planning. WICEM II, which took place in 1991, adopted sustainable development as its main axiom. The corporations agreed that there should be convergence, and not conflict, between economic development and environmental protection, and launched the Business Charter for Sustainable Development. In 1990, the Business Council for Sustainable Development (BCSD) was created under the chair of the Swiss industrialist Stephan Schmidheiny, personal friend of Maurice Strong (UNCED's Secretary General) and his special adviser for business and industry during the UNCED process. The BCSD was created as a group of 48 chief executive officers of corporations from all regions of the world, some of them with a rather negative environmental record, including Chevron, Volkswagen, Nissan, Nippon, Mitsubishi, Dow, Shell, CVRD, Aracruz, and Axel Johnson. The BCSD was closely involved in the preparation of the Conference, and, through Strong, had special access to UNCED's Secretariat. As a result, after Rio, corporations became "partners in dialogue," and their vision of sustainability became the dominant vision. According to Chatterjee and Finger (1994), corporations shaped the very way environment and development are being looked at: business and industry's worldview came out of Rio as the solution to the global environmental crisis and no longer as its cause.

In the words of the BCSD, "the cornerstone of sustainable development is a system of open, competitive markets in which prices are made to reflect costs of environmental as well as other resources. When viewed within the context of sustainable development, environmental concerns become not just a cost of doing business, but a potent source of competitive advantage. Enterprises that embrace the concept can effectively realize the advantages in more efficient processes, improvements in productivity, lower compliance costs, and new market opportunities." Thus, by creating competitive advantages, environmental concerns can provide corporations with new market opportunities and be the source of new profit. Finally, business sees the new era of global development as the era of market efficiency. "It is time for business to take the lead," says Schmidheiny; "change by business is less painful, more efficient, and cheaper for consumers, for governments, and for business themselves. By living up to its responsibilities, business will be able to shape a reasonable and appropriate path toward sustainable development". The ecological crisis perceived in fact by business not as a real crisis but rather as a set of adverse and controllable side-effects of development. Hence it is to be solved via increased efficiency which is to be achieved not through government regulation, but through open markets with a new concern for internalizing externalities.

Today, the BCSD has become the WBCSD (World Business Council for Sustainable Development), under the chair of Börn Stigsen. It now has 125 members representing companies such as British Petroleum, Ciba Geigy, Nestle, Monsanto and the Western Mining Corporation. The WCSD is said to have led industry input into the UN Commission for Sustainable Development and UNCED's 1997 review, revealing the emergence of corporate environmentalism as a driving force of global environmental management.

As stressed by Karliner (1997), after Rio, global corporate environmentalism has helped build a public image of transnational corporations as the world's responsible global citizens, setting the terms of the debate along lines favourable to their interests. In the process, corporate environmentalism has partially neutralized efforts - ranging from popular environmental movements to intergovernmental treaties and conventions - that pose a threat to their activities. While before Rio the environmental movement used the system to advance its goals, now the system has appropriated the environmental discourse and is using the environmental movement. This new strategy has meant increased efforts by corporations to increase cooperation with other environmental actors, in particular with the environmental movement. As noted by Bryant and Bailey (1997, 120), TNCs have sought to cultivate links with moderate NGOs in order to neutralize the threat posed to business from environmentalists. Actually, some NGOs today depend on TNCs for financial support. Stauber and Rampton (1995) observe that this process of funding NGOs and cooperating with them is part of a larger attempt to divide-and-conquer the NGO sector by winning support among moderate NGOs while attacking radical NGOs which campaign against TNCs' activities. Moderate NGOs and TNCs became partners in the international environmental establishment and now work together in the system of global environmental governance. From Rio 92 To New York 97: The Rise And Fall Of "Global Environmental Management" UNCED's Review Five Years after Rio Five years after Rio, as foreseen at UNCED, the review of UNCED's implementation culminated with the June 1997 New York Summit, often referred to as "Earth Summit II." Earth Summit II's official name is UNGASS, United Nations General Assembly Special Session. During UNGASS, five years of work of the Commission on Sustainable Development (CSD) were presented, including a report by the Secretary-General assessing the progress achieved in the implementation of Agenda 21 and recommendations for future action and priorities.

UNGASS was carried out at the highest level of political representation - Heads of State and Governments - and, as UNGASS itself said, aimed to "re-energize our commitment to further action on goals and objectives set out by the Rio Earth Summit." A new energy was indeed necessary: the main outcome of the meeting was the public recognition of the failure of international efforts to promote long-term sustainability. Yet it only adopted a document, the "Program for the Further Implementation of Agenda 21," and did not produce a political statement or binding commitments needed to reverse unsustainable trends. The text acknowledges that, five years after UNCED, the state of the global environment has continued to deteriorate, and reviews the situation in all areas of action.

It notes progress in institutional development, international consensus-building, public participation and private sector actions, which have allowed some countries to curb pollution and slow the rate of resource degradation. Yet, overall, trends are worsening, polluting emissions have increased, and marginal progress has been made in addressing unsustainable production and consumption patterns. Inadequate and unsafe water supplies are still aggravating health problems, the situation of fragile ecosystems is still deteriorating, and non-renewable resources are used at an unsustainable rate. Despite progress in material and energy efficiency, the report concludes that overall trends remain unsustainable. The document then reviews progress in all sectors and issues, inter alia, fresh water, oceans and seas, forests, energy, transport and atmosphere. Finally, it recommends means of implementation and adopts a program of work of the CSD for the next five years, with a commitment to ensure that the next comprehensive review of Agenda 21 in 2002 demonstrates greater measurable progress in achieving sustainable development.

Interestingly enough, all these trends are examined within the framework of economic globalization. The very assessment of progress made since UNCED starts by highlighting that the five years elapsed since then have been characterized by the accelerated globalization of interactions among countries in the areas of world trade, foreign direct investment and capital markets. The document recognizes the unevenness of the globalization process, stressing that marginalization and income inequality is increasing in some countries as well as within countries and that unemployment has worsened in many countries.

Yet it is believed that globalization presents new opportunities and challenges. The report notes that a limited number of developing countries have been able to take advantage of those trends, attracting large inflows of external private capital and experiencing significant export-led growth and acceleration of growth in per capita gross domestic product.

The view is thus that all countries could take advantage of the globalization trend. It is not perceived that only a few countries, due to specific conjunctural conditions, including interest rates and the monetary situation for example, can attract the volume of FDI necessary to feed the high growth rates praised in the document. The conceptual link with economic globalization appears as somehow flawed. It is not mentioned that significant export-led growth and the acceleration of growth in per capita GDP, if not controlled by an effective system of environmental protection, might be responsible for the worsening of overall trends for sustainable development.

In addition, though the text perceives unsustainable patterns of production and consumption as the major cause of continued deterioration of the global environment and observes that unsustainable patterns in the industrialized countries continue to aggravate the threats to the environment, only very vague actions and guidelines are adopted to change them, such as recommending the internalization of environmental costs, developing indicators, promoting efficiency, information, technology, and the role of business in shaping more sustainable patterns of consumption.

No binding commitment to deal effectively with consumption patterns or to establish sustainable production and consumption strategies has been adopted, and the role of actors who tend to promote unsustainable production and consumption patterns, such as business, is actually strengthened. As well as consumption and production patterns, another distorted linkage to structural economic conditions is made with the recognition that as a result of globalization, external factors have become critical in determining the success or failure of developing countries in their national efforts. It is rightly observed that environmental protection can only be promoted through a shift in the international economy and the establishment of a genuine partnership in order to achieve a more equitable global economy. Yet the idea is that the way to make all countries, in particular developing countries, benefit from globalization is through a combination of trade liberalization, economic development and environmental protection. It is believed that the international trading system should have the capacity to further integrate environmental considerations and enhance its contribution to sustainable development, without undermining its open, equitable and non-discriminatory character.

The text limits itself to recommendations to implement the Uruguay Round and promote trade liberalization. The reality of the present international trading system, a system which promotes discrimination against developing countries, consolidates global disparities and supports unsustainable practices not only in terms of consumption and production but also encouraging transport and pollution and shift from traditional cultures, is not seen as contradictory with the goal of long-term sustainability. With respect to transport, the text notes that the transport sector and mobility in general have an essential and positive role to play in economic and social development, and transportation needs will undoubtedly increase. It also observes that, in the future, transportation is expected to be the major driving force behind a growing world demand for energy. The document accepts that present trends are unsustainable, and adopted recommendations to make transport become more sustainable and mitigate its negative impacts. Yet the document fails to recognize the major cause of transport's expansion, namely, trade liberalization, which encourages production to relocate on the base of a traditional government subsidy to transports or allows for products originating at the other end of the world to be cheaper than products produced a few miles from the consumer. The fact that the whole globalization project is based on the continuity of cheap transport is not discussed.

Generally speaking, UNCED's review was critically received at all levels, being criticized both by diplomats, NGOs and by the press. Ambassador Razali Ismail of Malaysia noted that the compact achieved at Rio had eroded along with much of the high-profile attention to sustainable development generated by UNCED. And the Earth Negotiations Bulletin, a publication of the International Institute for Sustainable Development, noted that "in 1992 one could scarcely escape the news of UNCED and/or the environment in the media. This is not the case today... In international relations, perceptions are everything, and if UNGASS is ultimately billed as a non-event it will not bode well for the future of sustainable development or the UN in general during this critical time of its reform."

Most of the world's press was unanimous in condemning the failure of the New York Summit. The French newspaper Libération, for example, noted in its article "The Earth Summit goes round in circles" that the New York summit closed on an acknowledgment of impotence. Not only did the conference show the little progress accomplished in five years, it also failed to commit governments to significant concrete action and to provide means for implementing Agenda 21. No commitment was taken to achieve the goal of 0.7 % of GDP going to ODA, considered necessary to move towards sustainability. Development assistance today does not exceed 0.3% of GDP, on average, and, in the case of the United States, it was only 0.1 % in 1995.

The US was also the target of much criticism for failing to commit to effectively fighting global warming and to accept concrete reductions in levels of greenhouse gas emissions. At the end of the climate negotiations, no legally binding commitments to target and timetables emerged, and the conference only produced a watery compromise to seek satisfactory results at the then forthcoming Kyoto Conference on Climate Change, which took place in December 1997.

In short, on most major issues at stake, New York 1997 represented a backwards step in relation to UNCED's outcomes. NGOs speak of a scandalous betrayal of the Rio promises and of an utterly shameful outcome from Earth Summit II.

The reality is that the world has changed since Rio, and this change has a name: globalization. The Rio 1992 bargain was based on the commitment by developed countries to provide increased financial resources through ODA and technology transfer to help developing countries move towards sustainability. The implementation of UNCED's agreement was in a sense made dependent upon this aid. However, since Rio, ODA levels have been declining and the private sector has become the major agent of change. Government spending is being cut and state reforms are being carried out worldwide, often reducing not only ODA but also domestic environmental budgets. At UNGASS 1997, developing countries through the G77 tried to obtain a recommitment from the North to UNCED's bargain, including an increase in financial flows, technology transfer and an international economic system more favourable to developing countries. Yet today, as foreign investment replaces overseas development assistance in amount and frequency, UNCED's bargain seems politically outdated, and, as a result, its implementation appears highly jeopardized.

Finally, at the level of NGOs, the fracture among environmentalists is today stronger than five years ago. True, NGOs did lobby the CSD and try to influence the official negotiation process. Indeed, NGOs achieved unprecedented access to the intergovernmental process, with Greenpeace and the Third World Network being allowed to make speeches before the General Assembly. However, most of them had given up the idea of having a unified position on all environmental matters, and no "Global Forum II" was organized in New York, only an inappropriately named "Global Gathering" took place.

CHALLENGES FOR SOCIAL DEVELOPMENT

Some of the important social problems like poverty, ignorance, over-population and rural backwardness are of a general nature and, in varying degree, they are influenced by factors like squalor and bad housing, malnutrition and physical and mental ill-health, neglected childhood, family disorganization and a low standard of living.

For along time, society has remained apathetic to these conditions, but with the awakening of political consciousness and the enthusiasm of organizations and workers to improve social conditions, there is a possibility of developing programmes which could gradually remedy the present situation. The economic programmes of the Five Year Plan will mitigate these problems to some extent, but the gains of economic development have to be maintained and consolidated by well-conceived and organized social welfare programmes spread over the entire country. It is proposed to consider some of the more important problems of social welfare which need the special attention of both State and private welfare agencies.

The principal social welfare problems relate to women, children, youth, the family, under-privileged groups and social service. The social health of any community will depend a great deal upon the status, functions and responsibilities of the woman in the family and in the community. Social conditions should give to the woman opportunities for creative self-expression, so that she can make her full contribution towards the economic and social life of the community. Problems relating to health, maternity and child welfare, education and employment.

Some problems of women have to be dealt through social legislation, but other problems pertaining to health, social education, vocational training, and increased participation in social and cultural life, provision of shelter, and assistance to the handicapped or maladjusted call for programmes at the community level. As women have to fulfil heavy domestic and economic responsibilities, adequate attention has to be paid to the need for relaxation and recreation both in the homes as well as in the community. The welfare agencies have catered to some extent to the needs of the widow and the destitute woman, but the quality of the service rendered by them and the nature of their work needs to be surveyed.

Considering the numbers involved, the needs of children should receive much greater consideration than is commonly given to them. There is a growing demand for child health services and educational facilities. The standard of child welfare services in the country can be improved if the rate of increase in population is reduced. Problems relates to family planning, children's health, infant mortality, education, training and development have been discussed elsewhere in this report. Malnutrition is perhaps the major cause of ill-health and lack of proper growth of the child. The feeding of the child in the early years is the responsibility of the family, and is dependent upon economic conditions and traditional food habits. The nature and extent of malnutrition has to be determined, and resources have to be found to supplement and improve the diet of children through schools and community and child welfare agencies.

The problem of children's recreation and development outside educational institutions has received some attention during recent years, but play activities of children are considerably restricted in urban areas on account of the environmental conditions, lack of adequate space, and, to some extent, neglect of this vital need of the child by the family and the community. Not enough is known about the work of private agencies for the welfare of destitute and homeless children. The juvenile courts and children's aid societies have so far touched a fringe of the problem of children's welfare. Certain special aspects may be briefly mentioned. The existing facilities for handicapped and deficient children are far from adequate and suitable agencies have to be created. Hospitals provide treatment for polio, congenital deformities, fractures, bone disorders and other diseases, but there is a need to extend existing services and provide special institutions and care for disabled and crippled children.

At present deficient children attend educational institutions together with normal children and seldom receive treatment and special training to enable them to overcome their handicaps. The subject needs to be studied carefully. The problem of juvenile delinquency has already received considerable attention and many of the States have special legislation. Juvenile delinquency may often be the result of poverty and many offences may be traced to the connivance or support of adults. The youth constitute the most vital section of the community. In recent years, young people have had to face and have been increasingly conscious of problems such as inadequate educational facilities, unemployment, and lack of opportunity for social development, national service and leadership. The problems of health, education and employment of youth have been considered as aspects of national problems in these fields.

Social welfare is primarily concerned with the improvement of services provided for the benefit of youth by welfare agencies with the object of promoting development of character and training for citizenship and for physical, intellectual and moral fitness. It is necessary to encourage initiative among youth so that through their own organizations, they can develop programmes of youth welfare and national service. Ways must also be found to give opportunities to youth for active participation in constructive activity. Such training and experience will equip them for shouldering the responsibilities of leadership in different spheres of national life.

Traditionally, the family has been left largely to its own resources to deal with most of its problems, although in some cases it may be assisted by the larger community groups (such as caste) to which a family may belong. General problems relates to health, education and employment. Questions relating to status and rights, property, inheritance, etc., are the subject of social legislation.

The gradual break-up of the joint family and the emergence of the small family have increased its economic problems and burdens. Family responsibilities have now to be borne at a comparatively younger age by the head of the small family than happened in the joint family. This creates the need for greater guidance and assistance in dealing with family problems. The increasing complexity of the social situation and handicaps arising from physical disability, ailment or unemployment render it more difficult for the family to provide a sense of security to its members. This fact suggests a number of problems which, along with other problems such as divorce, desertion, and treatment of mal-adjusted members of the family, need to be studied carefully if welfare agencies are to develop suitable methods of treatment for guiding and assisting those in need.

There are a number of under-privileged communities such as the scheduled tribes, scheduled castes and other backward classes including criminal tribes. The problems of poverty, ill-health, and lack of opportunities for development affect them to a larger extent than many other sections of the society. The main problems to be considered under the description of social vice are prostitution, crime and delinquency, alcoholism, gambling and beggary. These problems have existed for a long period, although necessarily their nature and extent vary according to the prevailing social and economic conditions. Some of them have to be dealt with largely by local communities, and the approach and treatment have to be varied from place to place.

The character and magnitude of these problems of social defence have to be determined carefully before the value and efficacy of the existing agencies and programmes could be assessed. Social legislation deals with many of the social evils with a view to controlling and even eradicating them, but its actual implementation needs to be watched.

Among the practical problems to be resolved are the demarcations of the relative roles of State and private agencies, determination of the machinery of enforcement, estimation of the resources required, examination of methods, development of correct programmes, and creation of public opinion in favour of an objective and dispassionate approach to the problems of social vice. As the social structure becomes more complex, the State is called upon to play an increasing role in providing services for the welfare of the people. The Central Government, the various State Governments and local self-governing bodies, each in its own sphere, have to ensure that they have at least the minimum administrative machinery for dealing with social "problems. What form this machinery takes will depend on their particular circumstances and requirements, but it is certain that without the necessary machinery they will not be able to pursue their programmes.

Training for Social Work

The contribution which social services make will depend to a considerable extent upon personnel and leadership. A general understanding of the philosophy and history of social work, the structure and functions of society, the nature and extent of social problems, the methods and techniques of social work, and of the details of the programmes and how best their results may be assessed, will help improve the quality and efficacy of all services organized by State and private agencies. The training of social workers should of course include knowledge of conditions prevailing in fields in which they are to work, and social workers must possess the spirit of service and the character and energy to execute programmes despite handicaps and limitations and with such resources-as may be readily available.

There are several schools of social work in India and the setting up of some other institutions on similar lines is being contemplated in some of the States. There are important problems involved in these institutions which require specially qualified and experienced personnel, careful selection of candidates for training, special training for fields in which there is scope for employment, and adequate opportunities for field-work experience. Trained social workers are needed in large numbers for rural areas. It should be possible for the existing schools of social work to draw students from rural areas and to arrange for their training in the field in selected centres organized by rural welfare agencies. Universities and colleges in or near rural areas could also develop training programmes for rural development. Agricultural colleges could introduce intensive social welfare courses and field-work 'programmes as part of their curricula. Similar institutions with greater emphasis on social anthropology could be created in tribal' areas.

It is not possible for many voluntary organizations in the country to employ highly trained personnel for their ordinary programmes and activities. It is, therefore, necessary to arrange for training at the community level for field workers, instructors and supervisors. The existing schools of social work, specialized social service agencies, social welfare agencies functioning at the national and State level should provide opportunities for such training. Arrangements for ' in-service' training should also be made by the larger voluntary organizations which have worked in the field of social welfare for many years. Further, arrangements have to the made for the training of voluntary workers who will be needed in large numbers during the coming years. It is especially desirable that voluntary administrative and field personnel should be given some elementary training in social work.

The emergence of State social services and of large central organizations to deal with important social problems and the lack of opportunities for higher training in the social sciences within the country indicate the need in selected cases for training and study abroad in specialized fields. It is necessary that persons who go abroad for training should first have sufficient knowledge and experience of Indian conditions and problems.

5

Sustainable Rural Development in India

THE ONSLAUGHT OF COLONIALISM AND COMMERCIALIZATION

One of the first things that the East India Company did, by design, was the disruption of self-sufficient village economies, with interdependent social and agrarian relations, in India. The commercialization of agriculture and integration of village economy with the market, that served the interest of the British and middlemen, resulted in the breaking down of interdependence among the villagers and unemployment. The old benevolent privileged people of the villages became the collaborators of colonialism. The commercialization and breakdown of self-sufficient village economies impoverished the rural people and deteriorated their quality of life and livelihood. Their whole traditional system collapsed and got converted to the commercialized system; by integration with the market economy. The small landholders became the landless labourers.

Independence and After

The Panchayti Raj system was envisaged as reintroducing method of Gram Swaraj by empowering the people to deal with their own problems with the financial and technical support from the provincial and central government. It was the best possible effort at the time of inception. The ground realities in different states are mixed. There are states like Rajasthan, where Panchayats play major role in the democratic governance; and there are other states with varying degrees of empowerment.

Rural development and improving the quality of life in rural areas has been in force in our planning and developmental strategies all along. The numerous schemes that the Government of India has launched have had a sorry state of affairs in implementation.

The basic public services have not reached the rural areas especially the poor. The functioning of the systems is not encouraging the participation of all sections of the society. The transparency and accountability norms are hazy. The 'output' and 'outcome' of these schemes with waste expenditure have been poor.The IRDP that was planned with good thinking had inherent weaknesses in terms of monitoring and evaluation. It is too common to be further stressed that the benefits monitoring was very poor and there were large scale leakages and inefficiencies in the schemes; that even today are a cause of concern. Nonetheless, it was the first major initiative in the right direction.

The recent programmes like the Bharat Nirman, PMGSY, SSA, SJGRY, BRDF, NFFWP, NREGP, PMGSY, SGSY; and NRHM have been launched with substantial funding. The effectiveness of these programmes will depend on the implementation by convergence and synergy.

Several of these programmes may have the overlapping and complimentary components; if that can be synergized, it will result in huge administrative cost cutting and efficiency in implementation. One can not wait in the infinity to village people acquiring the management expertise. It would rather be more fulfilling to make them learn by doing in the simplest, yet accountable, possible ways.

Development Unit for the Villages not Block

In most provinces, panchayats are viewed as agencies and not the institution for development and change management. Both, the politicians and bureaucrats, have shown sustained reluctance to make the panchayats as institutions of self government. Even in the 'spheres of activities' their local presence is not truly recognized in the Constitution of India; leaving the scope of interpretation open. To some extent, this is made up under the various State Acts; and Panchayats (Extension to the Scheduled Areas) Act, 1996. However, they have to depend on the states and the centre, for financial resources.

The first thing that should be in place now, is the targeting the village panchayats as the nodal unit for the rural development and not the Block. The justification is simple-the development blocks have not proved effective agents of change in the villages! It is also true that the awareness and the outreach of the basic infrastructure and communication technology are now in a position to communicate directly with the villages. This will ensure the direct access to the problems areas and remedial action both ways. Much of the failure of development schemes can be attributed to 'top-down' planning and approach towards the people.

District or Block level planning can be only generic; while the specific plan for the development of villages should be the domain of the Panchayat. District and block level establishment should be used for conceptual issues, facilitation and capacity building of Panchayats. They should have nothing to do with funds allocation, disbursement and projects. If they have the expertise they can do evaluation and impact assessment, independently.

Long Channels and the Hierarchy

First thing to facilitate convergence and synergy is to remove the long channels of implementation, disbursement and monitoring. Each channel adds to delay, inefficiency and other problems.

It is too well known a fact to be discussed here that in the present system there is always a gap between the funds required in terms of both quantum and timing. Much of this problem is due to the complex system of percolating down of the funds, possible leakages; and delays at each level. The remedy lies in empowering, educating and energizing the Panchayat System and allowing funds to be remitted direct to the Panchayats. The hierarchy within the Panchayat system (District, Block etc. levels) leads to 'external' and political influence; as well as other factors that undermine the role and responsibilities of the Panchayats! The simple systems get complicated, to suit everyone's vested interests in the hierarchy. The decision makers must be the local villagers and the Panchayats.

Community Empowerment

In the rural areas, the outreach of the administration is not as much as it is in the urban areas. Besides, the villages have common properties and assets in several ways. These are not owned by any single family or person. The community empowerment will ensure the effective participation and indulgence of the people in the spheres of activities of the villages and development.

One way to ensure the community empowerment is to allow social audit of the major activities and programs of the government and the panchayats. This will ensure the representative groups being the part and parcel of the planning, implementation of development schemes and programs, monitoring and evaluation thereof; and the assets management.

EMPOWERMENT OF PANCHAYATS

Education and continuing education is essential for any organization. It is heartening to note that most states have prescribed educational qualification for Gram Pradhans. These educated headmen have the requisite local experience and are well conversant with the basic problems of the villages.

A *'Gram Pradhan'* is most accessible nodal point for every person in the village. In contrast with the MP's where there is no educational qualification required, the Gram Pradhans have had the benefit of education for furthering theirs views and ambitions in life.

The fact that, presently, Panchayats are more of a voluntary kind of institutions, it requires the attention of the government to make it as a career option for the villagers. Instead of spending millions on administering the schemes and programs at several levels, it would be prudent and more professional to make Panchayat jobs paid ones. The villagers and their representatives should have a major say in planning, project proposals and their implementation. In states where Revenue Police system is in place, they can be brought under the Panchayats for a better and effective system of dealing with law and order, revenue matters; and local disputes.

With such empowerment, the Panchayats will require to be educated on management of priority schemes, financing, and monitoring oversight mechanisms. The capacity building of the Panchayats and the villagers in several ways can be organized by making the jobs look like simple ones and focused to their own village; or nearby areas in cases of common facilities and shared resources. Training and capacity building will energize the Panchayats and make them effective implementers of the plans, schemes and projects of which they have been a party. The whole system of Panchayat in the village will have to be energized by involving all sections of the population in the villages.

Low-Key Ombudsman

The monitoring of the effectiveness depends on good management information and accounts keeping. For purpose of accounts keeping regular position of an accountant can be created. The major concern that we may still have is to keep a proper vigil and watch on the speed and transparency.

I would strongly suggest that there can be a low-key ombudsman type arrangement by hiring the services of retired teachers, retired army officers or any other available eminent people with proven integrity and patriotic feelings. I will never have doubt about these qualities amongst retired teacher or army personnel who generally work for devoted cause and commitment.

e-Governance

Records keeping and retrieval of information for the land records, common properties, demographic information, assets etc. is generally poor in cases of all the villages. Sometimes, the poor and illiterate people do not even have access to such information. There is a growing need for use of information technology for possible solutions in the area. Wherever possible, e-Governance should be supported.

It is a myth that right now it is too early for the e-Governance in the rural areas. There are the problems of skills; but the same can be enabled by hiring people or associating the NGO's.

A case in example is the Ernakulam district panchayat in Kerala, who has taken up a unique model of e-governance to establish information networks as part of their plan projects; with a view to strengthen the decentralised planning system. However, due care has to be given to the fact that most rural population will require handholding and assistance in the new scenario. The stems in place will have to very responsive to the concerns and queries of the villagers.

Convergence of Schemes/Programs

The convergence of all ministry specific schemes in the Panchayat's Plan for implementation should have an overriding provision to reprioritize the schemes during the financial year by utilizing the available resources, for local modified needs. Here we are presuming the oversight and monitoring mechanisms enabled in the system.

The rural knowledge bank available with the senior citizen will come handy to decide as to what is more important for the village as a whole at any given time. In any case, the canalizing funds directly through the Panchayats will result in synergy and economy by containing the wasteful expenditure incurred in the name of administering the grants for the schemes. The resultant reduction in government jobs can be utilised by redeployment of skilled staff in the Panchayats; and as independent consultants operating privately. The system must ensure that there are *ex ante* targets fixed for *ex post* results and benefits monitoring, quantified in both physical and financial terms.

Local modifications in Schemes/Programs

Local modifications may be allowed to cater to the needs of the varying geographic conditions, allowing for the cultural factors; and better value for money derived. The housing or the water supply projects can never have the same specifications in the mountain areas of North-East or the middle Himalayan states. The development of local specific and user friendly technology in the projects will help achieve the goals better than any standard scheme or projects for all states.

Social Solutions and Holism

The dynamics of interrelationship of land holding, agricultural growth, employment; and poverty have to be understood in holistic way and strategies have to be devised to tackle them in tandem. People–centric and people oriented methods will entail the participation of the civil society, self-help groups, stakeholders, NGO's and other institutions to help achieve sustainable rural development with a clear vision and focus.

Fragmented and piecemeal efforts will only remain attempts without being translated into outcomes; and a burden on public exchequer.

Several of the complex problems do not require economic or political solutions; but the social solutions.

We seem to be going overboard in bringing in more and more economic and political solutions, at the cost of neglecting the social solutions. The social changes are slower than even the slow process; but can be brought out only by tackling the social elements in association with the other measures and variables. The core areas of thinking shall be the holism of rural development:

"But holism requires much more – it must include the sense of all inclusive totality. Interestingly, this emphasis on all inclusive totality places at the disposal of less developed societies, an inexpensive volume of relevant R&D at the highest level of the corporate world which is struggling hard on the same problem of development – how to change the ongoing in order to survive in the world of future.

Why? Complex dynamic systems, whether physical, biological or societal, have the same underlying internal dynamics and exhibit common emerging patterns. All that is needed is the *cultivation and development of re-investment skills."*

Financial Management

Financial management and records keeping of any system is very important. We cannot expect the local bodies to perfect them without any regular support staff and system. Right from planning to outcomes reporting, skilled staffs are required. As said, elsewhere in this paper the management of Panchayats should be a career option for the villagers; and the requisite internal expertise has to be available in the organization. Panchayats must have their own Budget and some sources of Income, by having powers of levy taxes, user charges etc. in the local sphere of activities. This will automatically involve everyone in the village administration, through the concerns of the tax payers and the users of services. While the bulk of the sources of expending will continue to come from the State and Center's funds, the funds raising powers of the village panchayats will make them more aware of the areas that can be governed more effectively at the village levels. The maintenance of various existing and future assets that always starve for funds, will also find a new source for this purpose.

The large paraphernalia of the multi-tier system will provide the surplus manpower for redeploying in the Panchayats. It will in fact result into savings for the government in several ways. To begin with the panchayats may keep their cash based accounts on the 'single entry system', that reveals much more information on the transactions than the double entry or accrual system. The systems have to be made simple and transparent, unlike the complicated 8 Pallavi Dhyani (2003), Technological nurseries for rural industrialisation an d entrepreneurship data that is sought under the development schemes at present. In a fully accountable and transparent system, enabled with internal controls and audit, micro-management must be left to the Panchayats and only monitoring and evaluation should be the concern of the State and Central governments.

Internal Audit

An institutionalized mechanism of internal audit by independent agency/ NGO or even a kind of social audit system (of course making it simple) can be put in place. Most government institutions in the provincial and district administration also require having similar arrangement to make them accountable for their decisions.

Learning from Others

The empowerment of the rural local bodies will ultimately ensure that the 'ruralliving' is more prestigious than living in urban areas. In most European countries, living in the countryside is considered more prestigious than living in cities. Countries like U.K., Austria, Switzerland, and France are telling examples of that. Why they have such a situation is due to the comparable or better schools, hospitals, playgrounds, water supply, roads and general environment. In Alpine countries of Europe, there have been trends of reverse migration, with the upgraded rural livelihood.

Reinforcement of national programs and activities through decentralized decisionmaking and capacity-building has been undertaken by most of the countries in the Eastern Europe. Activities range from development of farmer support groups to the establishment of central and regional development councils, as in Hungary, which are responsible for regional planning, priority setting and allocation of financial resources for various regional development projects. Through the Estonian Village Movement, rural communities have been active in appealing to the Estonian Government for assistance in rural development issues. The first Rural Assembly for Villages was held in Estonia in 1996 and concluded with the formation of the Estonian Movement of Villages and Small Towns (KODUKANT), a nongovernmental organization.

The international community recognizes that concerns about the impacts of globalisation and the free market, nutrition, supply of safe drinking water and energy, and improvements in education and health.

Government measures would go a long way to solving problems of poverty, but improved coordination, cooperation, and targeting of investments was needed. The unique characteristics could be exploited to capture niche markets for products and services such as organic food products and eco-tourism.

The availability of such amenities and infrastructure enables the rural tourism to compete strongly with the city tourism. One the rural tourism catches up, the rural people will find jobs in their backyard of their ancestral village home; the pain of cutting the umbilical cord will also be minimized. This sounds very simplistic but requires a lot of investment, will and long term planning. Cities as 'Engines of Growth' will provide jobs and income; but certainly not rural development as most investment analysts think.

The road to emulating the European countries is not straight one. We have to have the town planning, drainage system, roads, schools, hospitals, safe drinking water, access from the cities, communication, etc. will provide the basic amenities comparable to the cities. The emerging concepts of 'Provision of Urban Amenities in Rural Areas' (PURA), have to be translated into reality. An intensive R & D is required to be carried out in agriculture, materials and renewable energy technologies.

TRENDS IN VOLUNTARISM IN RURAL DEVELOPMENT IN INDIA

My emphasis is much more on voluntarism than on voluntary organizations as instruments of social change. I do not disregard the niches that market forces and state and public agencies leave unfilled, but I argue that these niches can be fcilled not only by their third sector or voluntary organizations but by the "developmental deviants" or "entrepreneurs" or "volunteers." These volunteers, while remaining in the mainstream public or market organizations, can create new alignments between social needs and institutional support. The excessive attention on voluntary organizations by aid agencies seems misplaced insofar as these agencies almost completely neglect the DVs.

By supporting only NGOs, agencies reduce pressure on public and market agencies for reform and self-renewal. NGOs led by managers or leaders who are often from an urban context, by their own creativity, suppress or fail to nurture the creativity of the local disadvantaged. Social change thus becomes more and more dependent on external leaders. Rural development as par tof social change is defined here as a process of expanding the decision-making horizon and extending the time frame for appraising investment and consumption choices by rural disadvantaged people collectively, and not necessarily at the village level but at even higher levels of aggregation.

Sustainable processes will require correspondence between people's *access* to resources, *ability* to convert access into investments (that is, skills for using resources), and *assurance* of future returns from present investment (vertical assurance) and about others' behaviour vis-à-vis one's own (horizontal assurance or collective rationality). The changes in the network of access, ability, and assurance for DVs and the people have to be achieved simultaneously.

Voluntarism may affect any one or more subsets of the developmental triangle of access, assurance, and ability of the people and thus may remain restricted in its impact. The propositions that follow deal with the way that voluntarism has been related to the process of social change in India. Given the range of experiences, it is indeed a synoptic account.

Process of Voluntarism

Voluntarism triggered by a natural crisis such as flood, drought, or cyclone may legitimize the entry of outsiders in a given region, but depending upon the mobilization process, NGOs that emerge in response to such crises often diversify into other areas of social development and remain community oriented rather than class oriented.

Several church-based NGOs came into existence when international aid agencies offered relief at the time of the Bihar famine in the 1960s. Most of the relief was in the form of consumables such as foodstuffs, clothes, and medicine. The organizational structure for the distribution of this aid was different from the structure for managing durable assets such as rigs for drilling wells, transportation, and buildings. The move from relief to reconstruction attracted many young people. Instead of going back to pursue their professional careers, they remained behind to organize people, manage food for work programs, drill wells, or provide health and education facilities.

Many aid agencies sought legitimacy through relief but subsequently indulged in other interventions. The reaction of state agencies was to incorporate such volunteers or voluntary organizations as appendages of public relief and development programs. Such incorporation also took place in many NGOs, which came into the picture much later. An interesting feature of these organizations was that having begun with a community approach (relief was needed by all), they continued to use an eclectic approach to development.

Social conflicts were merely noted by some and participated in by others. The institutionalization of voluntarism in intermediary support or funding organizations or grass-roots organizations gave a techno managerial start to the intervention strategies. A negative feature of such aid was that in regions prone to frequent natural calamities, people started losing their self-help initiative. State relfief in the form of employment or food was not linked with a mobilization of voluntarism among the people. Dependency so created made the task of many radical NGOs even more difficult. People could not understand why mobilization around a radical ideology should be a reason for forging immediate material benefits.

Voluntarism triggered by man-made disasters such as the Bhopal tragedy can *get caught in the dilemma of legitimizing the state's indifference by becoming part of urgent relief and rehabilitation vis-à-vis questioning the basis of the tragedy and the complicity of the state in its consequences.*

Ravi Rajan (1988), while analyzing rehabilitation and voluntarism, observed four distinct styles: (1) intervening organization took on the provision of relief and rehabilitation as its primary task, became dependent on the government, and with the diminution in the governments' own commitment to the cause, soon collapsed; (2) volunteers served as "conscience keepers," pursuing change through systematic research reports; (3) trade union activists demanded charge of the industrial plant to provide employment through alternative use of plant and machinery; and (4) perhaps the most significant strategy by volunteers was to reject the idea of voluntarism as propounded by the state. Rather, voluntarism was redefined to include sustained mobilization, the struggle for better relief, access to medical data, questioning the scretiveness of the part of the government, legal activism, and questioning the right of the government to give such a low priority to the life of the poor. Voluntarism of this nature is difficult to mobilize in backward rural areas given the dispersed nature of settlements and weak social articulation, low media attention, and poor networking among interventionists.

Voluntarism as manifested in the 1960s by a protest against agrarian disparities (in the form of a violent leftist movement, known as the naxalite movement) and by social reconstruction (initiatives by students, professionals in the mainstream organizations, and voluntary organizations) has undergone a sea of change in the wake of recent economic liberalization.

Radical groups using violent means of social change have sought support essentially from Maoist philosophy. After the Chinese aggression in 1962, covert support to these groups increased, and income disparities intensified after the first phase of the green revolution. Technological change had provided the spur for a large number of young people, particularly from West Bengal and Andhra Pradesh, to plunge into the field of violent social change. The attempt was to annihilate rich farmers and other symbols of perceived oppressive classes or those considered class enemies.

Another stream of volunteers who entered the field of rural development came with innovative ideas for providing relief during the 1964-1966 drought in different parts of the country. These volunteers became crucial instruments of social dynamics. The war in 1965 with a neighboring country led to a slowing down of U.S. aid to India. The search for indigenous alternatives became intense, and the legitimacy of voluntarism increased.

The period between 1966 and 1972 was full of economic crises. The economic environment in the preceding decade had been aimed at the closure of the Indian economy through import substitution. Droughts, wars (1965, 1971), devaluation of currency, and inflationary pressure created an environment of social unrest in the organized and unorganized sectors. Death from starvation was supposed to have been eliminated (almost) after the drought of 1965-1967. Maharashtra started an employment guarantee scheme during the drought of 1972. In the wake of large-scale violence in 1966 and 1967 by left-wing radical groups, the report of a confidential inquiry committee by a committed civil servant (Appu) set up by the Home Ministry argued for an immediate thrust toward target group oriented programs of rural development suited to location or ecology and class-specific need. The Small and Marginal Farmer and Agricultural Labourer Development agencies, the Drought-Prone Area Programme, and the Tribal Development and Hill Area

Development plans followed. Decentralized development in the policy was accompanied by greater political centralization from 1970 to 1977. A movement based on Gandhian values that called for total social revolution was spearheaded by Jaya Prakash Narayan in 1973 and 1974. It attracted a large number of young people, particularly in Gujarat, Bihar, and Maharashtra, and many of these young people continued with voluntary work.

The government declared a state of emergency from 1975 until 1977, after a prolonged railway strike, and even urban people realized for the first time the implications of a non-democratic coercive state. Voluntarism was also sought out as a sign of despicable deviance. People has the option of being incorporated into the repressive state structure or being jailed or victimized. The post-1977 phase of change in political continuity through the single-party rule brought many Gandhians committed to decentralized development into the mainstream. Tax concessions for voluntary initiatives by commercial companies were introduced for the first time by the Janata government in 1978, and many innovative organizations came into being. A number of developmental volunteers who worked in commercial organizations found this an opportunity for exploring new organizational space. Some misused this option but many did not.

For the first time, professionals and young activists were offered competitive salaries in addition to autonomy for work unheard of in mainstream organization by and large. These events were also accompanied by a change in the policy of international aid agencies, which started shifting from funding better implementation of government programs bureaucratically to better implementation by NGOs. It was unfortunate that creative avenues in the NGOs. It was unfortunate that creative avenues in the NGOs got generally fossilized because of their proximity to the state and their participation in implementing standardized programs.

A change of government in 1980 and the restoration of rule by the Congress party led to the expected withdrawal of tax concessions; the centralization of voluntarism (companies could contribute to the Prime Minister's fund for rural development and seek fresh grants from it for action programmes); the halting of the direct transfer of funds from a commecial balancel sheet to the social (less easy to account) balance sheet; the standardization of developmental programs such as the Integrated Rural Development Programme (IRDP); the withdrawal of higher allocations to the IRDP for backward areas and putting them on par with other areas; and the merger of earlier adaptive or responsive programs into a standardized IRDP, with credit-linked subsidy as the dominant mode of relationship. Another interesting development was the return of naxalite (radical Maoist leftists) underground workers to the mainstream was non-violent but articulate strategists of social change. For the first time, several ex-naxalites sought election in 1977 and some were elected.

The social space for alternative development was filled by volunteers with vlarying backgrounds; ex-radicals; liberal or social democrats who were dissatisfied with the workings of the state and wanted to influence the distribution of resources; enthusiastic urban activists who after looking for a career, failed to get one and returned to a mainstream profession rather quickly; young professionals with technical or other disciplinary backgrounds who launched action-research projects or supported other professional groups; retired civil servants, ex-Gandhians, lawyers, and so on, who formed independently or with the support of aid agencies large NGOs; and quasi-state organizations promoted to provide technical, financial, marketing, or other support to NGOs, artisans, and other beneficiaries of state-sponsored developmental programs.

At the time when social space for volunteers was widening, opportunities for career growth in mainstream organizations also began to increase. The first phase involved the growth of the banking sector after nationalization in 1969. A large number of bright young men and women with backgrounds in science, the humanities, or engineering joined banks, insurance corporations, and other such systems. "Brain drift" as opposed to "brain drain" took a heavy toll by depriving academic disciplines of bright students and luring some professionals on the margin away from other direct social development systems. The post-1980 boom in the consumer goods industry and the continued growth of banking and other public and private ventures further increased the flow of young people toward such careers. The opportunity cost of those who chose to work in NGOs did indeed increase. The question we want to address next is "What are the processes by which voluntarism in mainstream organizations can complement the efforts of NGOs not merely in bringing about social change on the micro-level but also in influencing public policy in favour of the disadvantaged?"

Implications for Action and Research

Generating extra-organizational space for developmental volunteers within mainstream organizations is a necessary condition for sustainable social development. One study on bank and NGO cooperation for poverty alleviation in backward regions noted that there was no NGO working in the fifty most backward sub-regions of Gujarat. State organizations like the National

Bank could not gain credibility in supporting NGOs if they did not provide the opportunity for exploration and experimentation to volunteers within their system. NGOs often did not recover even the operating costs of many services from people. In the process, such NGOs remained perpetually dependent upon aid agencies. Moreover, accountability of the NGOs in regard to the poor was so low that most NGOs did not aim at inducting poor people into their own management structures.

One nationalized bank invited its clerks to volunteer for two years in a village development program in an area of their choice without any loss of seniority upon their return. This triggered numerous innovative experiments by DVs.

The hands of DVs in technology generation, adaptation, and diffusion system working on unpopular problems of larger social concern needed just as much attention. How can professionals who disregard professional rewards and devote attention to such problems but cannot put pressure for reform on their own organizations be sustained? Empowering them will require recognition of their voluntarism by a body of concerned scholars and activists. No national award has been given to date to any bank officer for initiating innovative schemes. So much so that about ten million rupees for new innovative schemes for rural development provided at the national level remained unspent because no system existed for identifying and recognizing DVs within the mainstream system.

Can the capacity of urban people to manage their own affairs be provided by urban volunteers who come from very different cultures? The poor do not cooperate with developmental organizations because they are not recognized as possessing any *richness* in terms of their cultural and moral fiber. The findings of our research are seldom shared with those from whom we collect data. Involving the rural poor as co-researchers of social phenomena, building upon cultural roots of voluntarism, and showing respect for common property institutions can change this situation. Acknowledging local initiatives can spur their transformation into innovations. Documenting people's knowledge and identifying the scientific merit of some sustainable resource management alternatives can rekindle people's experimental ethic.

Institution building requires the dispensability of external leadership, the recognition of an inverse relationship between status and skills, and the discrediting of values that generate helplessness. "Lateral learning" among developmental volunteers and NGOs can be triggered to provide empirical basis for building a "theory in and of" action.

Concepts of voluntarism such as zakat among Muslims, gupt dan (anonymous charity) among Hindus, Kar Seva (voluntary labour for the common good) among Sikhs, and so on, are examples of the positive bases on which different religions build organic institutions. Different languages have words like *andi* (Haryanavi) and *dhuni* (Hindi), implying a person obsessed with ideas generally for the social good. Why has appreciation for this trait vanished? Anonymous voluntarism, a unique and long-standing tradition of the east, has been absorbed by voluntary organizations that believe that voluntarism can only exist in their types of organizations. This vision is limited, because it denies the possibility of institutionalizing culture throughout a full range of institutions, not just voluntary organizations.

Finally, neither NGOs nor the developmental volunteers can succeed unless those long-ingrained values that inhibit change among rural poor people are brought into question. Voluntarism in rural development in India has not been accompanied by pressure for policy change except in regard to environmental issues. Often action at the local level has not been linked with lobbying at the macro level. Recognizing that the state and markets perform better if kept under constant check, developmental volunteers within the organizations will have to serve a sort of "insurgent" function so as to align, anonymously, with grass-root activists, NGOs, and professionals. International agencies can strengthen local social change by broadening local ideas and innovations into global thinking and by providing global space for developmental volunteers to validate their hypotheses. Right Livelihood awards constitute one such source of international recognition. If the rural poor of India could communicate with the homeless in America, surely the cultures of deprivation would provide the basis for collection action. Social innovators and DVs around the world are struggling for similar space in a society where one does not have to go through a phase of unbridled accumulation followed by guilt, charity, and benevolence for the have-nots. Sustainability in nature and society requires players, whistle blowers, spectator rules, and creative chaos. DVs are arguing that the losers in a game should not lose the right to play on the same field again. Asking them to play only on separate fields (in the form of volags) will eventually rob the game of the chaotic waves of sorrow and joy. Should we let it convert the spectators into warriors?

INDIA: COMPLETION EVALUATION OF ORISSA DEVELOPMENT PROJECT

The tribal communities are among the most underprivileged people of India. They are severely impacted by poverty and its manifestations, such as low levels of literacy and health care, hunger and malnutrition. They suffer social and political marginalisation and remain vulnerable to exploitation by the more powerful.

In order to redress this situation to alleviate the suffering of tribal communities and to remove obstacles to their development, further amendments have recently been introduced in the Indian Constitution. This laudable act should create an enabling environment for promoting the interests of the tribal communities and for harnessing their untapped potential for overall social and economic development. Towards this end, even more action-oriented programmes and projects are required which specifically address the constraints and opportunities for sustainable livelihood and empowerment of the tribal communities.

The Orissa Tribal Development Project (OTDP) was the first of its kind financed by IFAD in India specifically aimed at improving the livelihoods of the tribal people.

Its design may not have benefited from the evolution in developmental approaches and policies since its inception ten years ago; yet it has made a noteworthy contribution to highlighting the need for concerted action for ameliorating the social and economic conditions of the tribal communities. It has also been instrumental in refining IFAD's understanding and approach to its work in the sector, and since the Orissa project the Fund has financed four additional projects specifically targeted at tribal people in India.

IFAD's Office of Evaluation and Studies (OE) undertook a completion evaluation of the OTDP in November – December 1998. This was the only evaluation IFAD conducted of the project, and as such the evaluation process generated an interesting and diverse range of knowledge that we hope will contribute to improving future programmes, policies and interventions in the sector.

OE takes this opportunity to convey its appreciation to all those involved in the evaluation exercise, including the tribal people who generously offered their time and attention, the government and its officials who facilitated the evaluation and gave useful guidance, and to the representatives of the grassroots institutions who provided invaluable data and background information. A special thanks is due to our core partners in the evaluation process, including the Government of Orissa, the Delhi Directorate of WFP (the co-financier of the project), and our colleagues in the Asia & the Pacific Division in IFAD for their support and contribution to the learning generated through this evaluation exercise.

SEVEN LESSONS LEARNED—HUMAN RESOURCES DEVELOPMENT

The project had a component for Human Resources Development (HRD), which was to be implemented by a qualified NGO. HRD included the building of awareness and self-reliant capabilities among the tribal community in areas such as environmental management, trading practices, money lending, legal and land rights and social and economic development opportunities. However, the OTDP is a classic example of a development intervention in which the "hardware" side of development was given far more weight than the "software" side, both during design and implementation.

First and foremost, this is reflected in the allocation of financial resources for HRD activities: a mere six percent of total project resources were earmarked for HRD. Even the project management component was allocated more resources (eight percent).

Secondly, right from the design stage the project had an over-bearing faith in technology. It attached over-riding importance to establishing physical infrastructure, agriculture development, land allocation and related hardware activities as the motor of the development process. Consequently, as mentioned above, the project's funds were allocated mainly in favour of technical and infrastructure activities. This trend continued during implementation, as is, for example, illustrated by the limited efforts made by the project to identify a suitably qualified NGO to replace Agragamee. Instead, OTDP directly implemented HRD activities without having the required experience and profile. This reflected the failure of the project to recognise the prime importance and contribution of the HRD aspect to the overall project objectives and the related development process. In fact, the evaluation believes that limited success in HRD was a principle cause of the shortcomings of the project, since, for example, the target group did not feel "included" and sufficiently integrated in the project.

What the experience here clearly illustrates is that the software side of the development activities such as building participation and training, empowerment and education of the target populations are essential for the success, impact and sustainability of any project or programme of this nature.

The OTDP attempted to promote the well being of the tribal people by following a traditional integrated approach to development through investment in agriculture production, utilisation of natural resources, rural infrastructure, and land survey and settlement. However, the need to "prepare" the development process in the area of social mobilisation and social development prior to launching full-blown productive investment activities cannot be overstressed. In this way projects and programmes will become "people-centred", thereby instilling a sense of trust and confidence in the beneficiaries towards the project and leading to the requisite sensitisation of the target group as owners and stakeholders. Only thus would they benefit as intended and, equally importantly, be motivated to contribute effectively and willingly to the success of the project, both during implementation and beyond. The importance of HRD for project authorities is also a dimension to consider. Personnel responsible for project management and implementation need to be sensitised about the project's objectives and the importance of building grassroots capabilities, and to generate concern and commitment in them towards the intervention.

Knowledge of the Rural People

The OTDP highlighted the role and importance of the knowledge of the rural people in the process of the design and implementation of development projects and programmes. Several examples from the project's evaluation may appropriately illustrate the above. For instance, when the evaluation mission visited a particular village in the project area, it noted excessive silting and damage to some of the Water Harvesting Structures (WHSs). This was surprising given that the WHSs had been developed only a few years earlier.

Following discussions with the tribal people, it was learnt that they had impressed upon the concerned engineer for an alternative site for constructing the WHS. This was based on their intimate experience and knowledge of local climatic conditions, water flows and agro-ecological conditions of the terrain indicated for the WHS. However, their suggestions were disregarded and the decision to maintain the site selected a priori was taken by the project authorities, upon the recommendation of the engineer. Another example is the selection by OTDP of coffee as a crop for development on the lands allocated to the tribal people. Very little, if any, coffee was planted during the project by the target group, as this crop is traditionally not favoured due to cultural and economic reasons by the tribal people in Kashipur block (i.e. the project area).

The OTDP evaluation has illustrated the crucial importance to the development process of the knowledge of the tribal people. The tribal people have survived in their environs for centuries without huge amounts of money being spent on their development by governments or others, and over the years they have developed and refined valuable knowledge and problem-solving strategies not only concerning agriculture, farming systems, natural resources management and biodiversity, but also in the areas of health and education, and social organisation and mobilisation (e.g. by forming associations and groups for credit and savings, water use, home economics). Their knowledge and experiences need to be tapped more systematically, and blended accordingly with "modern" technical knowledge suitable to the context and environment in which the tribal people live and operate.

By making use of tribal peoples' knowledge and experiences, development activities and projects will be relatively easier to implement, and their impact is bound to be far greater. The evaluation has thus drawn attention to: (i) the need to avoid a one-way, top-down approach to knowledge and technology transfer; and (ii) the value and intensity of untapped knowledge of the rural poor, such as the tribal people in Orissa, which, if used appropriately, can be a useful catalyst in accelerating the development process.

It should also be noted that design missions ought to carefully analyse all aspects related to the tribal population's practices before designing any form of development intervention. In this regard, among other areas, attention should be given to their cropping patterns and intensity, adoption of technologies, labour force availability. These issues could be analysed by undertaking special studies at the design phase, and kept uppermost during implementation so as to make the best use of locally existing skills and know-how and to ensure that the evolving requirements of the targeted populations are constantly taken into consideration and catered to.

Land Rights and Tribal People

The OTDP implemented a component for land surveying and settlement, which resulted in the allocation and distribution of dongar (hill) lands to the tribal population in Kashipur block. Until recently, the population in the block were making use of the dongar lands in an informal manner, as and where possible, for agricultural purposes (mainly shifting cultivation). However, access to these lands was not secure and consistent; and land productivity was low.

The benefits of legally secure land tenure are evident from the OTDP experience. From an agricultural point of view, the allocation of parcels of lands and the provision of land titles has reduced shifting cultivation practices, which has consequently promoted relatively sound environmental management practices and helped restore agro-ecological balance. It has also substantially increased the productivity of these lands. From a socioeconomic point of view, the project illustrated that when land titles are registered in the names of both spouses, the social and economic status of women is enhanced, providing them with greater security, confidence and independence. It has also provided them with more opportunities for income-generation through activities such as vegetable gardening and small livestock rearing. Overall, then, the ownership of even a tiny piece of land has improved the economic conditions of those concerned. It may be noted that a somewhat better level of literacy was also observed among the beneficiaries.

The Orissa evaluation mission witnessed another feature of prime concern in the land reform process under the project, which concerned the implication on community ownership and management of forest lands and water. Such and other common property resources are particularly important for tribal people, since they derive a large part of their income and nutrition through the processing, consumption and sale of minor forest products. But, as mentioned above, much of the community owned land was individualised, and land titles were recorded jointly in the names of both spouses. While this step was welcome, the package of privatisation of property upset the existing social security for the tribals. Now smallholders in distress could either lease their land to moneylenders and others, or be mobilised by the development process for eviction, settling for cash compensation for their piece of land. In fact, the latter was observed in several areas in Orissa as a result of the development of alumina projects funded by multi-nationals. Hence, such land reform programmes need to be accompanied by opportunities for employment and marketing, access to rural financial services, and institutional support. Provision of such services will contribute to fully unleashing the potential and productivity of the targeted populations.

With regard to the protection of the tribals rights over community forests and other lands, following the central legislation introduced in 1996, the Gram Sabha (Village Assemblies) in the tribal areas has been entrusted to protect the community rights over community land and forest. The OTDP has preceded this Panchayati Raj enactment, and unfortunately this power is not being exercised by the Panchayats in many tribal areas (in Kashipur itself, the GOO has notified the transfer of tribal land to the alumina project companies without consulting the local Panchayats). Moreover, some community lands have also been encroached upon by the rich and powerful. Therefore, projects and programmes promoting land reform should simultaneously pay sufficient emphasis on the management and conservation of community resources such as forest lands. Furthermore, the communal rights of the tribal people over such areas should be safeguarded, as they form an integral part of the overall production system of the tribal populations.

Project Management

A special project management unit (PMU) was set-up to implement project activities. The PMU was headed by a project manager, who had a team of technical specialists on secondment from the relevant line departments of the GOO. The relatively poor performance of the project management adversely affected project implementation.

This was due to several reasons including the frequent changes of project managers, a top-down approach to implementation with limited participation of the various stakeholders, lack of an appropriate M&E system, frequent key staff vacancies and poor coordination amongst the PMU technical staff originating from the various line departments. With regard to the latter, a dimension to be noted is that some OTDP technical staff were stationed in Bhubaneswar and others in Raygada, which did not facilitate coordination.

OTDP project management would have benefited from an alternative approach. The frequent transfers (and changes) of project managers created "instability" in the management of the project.

There was even a disincentive for project mangers to perform well, since high performers may have been requested to continue on the job whose location was remote and isolated, and on which limited financial incentives were available. As long as project designers maintain the management function within the existing government structures and authority, it is not easy to request governments to alter their practices and policies in this regard.

In fact, not only IFAD but also other large donor-funded projects and programmes in India are subject to similar government procedures with respect to staff appointments and transfers. Another related issue is the process of selection and assignment of staff from technical government departments to the project. The OTDP suffered from the lack of a transparent process of staff allocation, some of whom were not adequately equipped technically to discharge their duties effectively. Project management was also not sufficiently prepared for a participatory approach in implementation, and this marginalised the various stakeholders, principally the NGO Agragamee and the beneficiaries themselves.

Continuity, competence and stability of project management are key ingredients for the success of the project, and, moreover, effectiveness of implementation is enhanced when managers and other staff appointments are based on merit and conducted in a transparent manner. Additionally, continuity for a period of three to five years in project management is necessary to build relationships, trust and cooperation among the various project stakeholders.

Project managers should be given appropriate incentives and special allowances to perform and remain committed to the job, especially in those duty stations which are remote and where the requisite social services are lacking. Project designers should also explore alternative and innovative approaches to project management. For example, implementation could be undertaken through established community-based organisations, civil society, the private sector, or through other competent institutions or bodies which are not under the command of government structures. Such arrangements would also shield the project management units from political and bureaucratic pressures of one type or another. Additionally, it is the responsibility of the project management team to foster a spirit of consultation and dialogue among the various project partners in the best interest of the project. In this regard, project managers and other concerned staff should be provided on-the-job training at entry. They should also be required to frequently visit the project area to remain constantly informed about field-level issues and requirements. To this end, adequate provision must be made in project costs to cover training and travel-related expenses. The involvement of reliable and qualified NGOs in selected aspects of project management should be encouraged.

To enhance chances of project sustainability, a minimum project management facility should be maintained after project closure for post-project activities. A provision for this should be made in the project design.

ROLE OF THE CO-OPERATING INSTITUTION

UNOPS as the co-operating institution for the project was responsible for supervision and loan administration. With regard to the latter, UNOPS ensured that the borrower complied with the covenants of the loan agreement, with the exception of the non-compliance related to the systematic preparation and submission of Audit Reports. Internal audits were conducted by the Orissa Tribal Welfare Department, but these did not measure up to the required standard. No external audit was conducted, as required by the loan agreement. In terms of supervision missions (SMs), these were conducted in a timely manner and UNOPS submitted supervision reports to IFAD promptly according to agreed time-frames. In fact, UNOPS mounted two SMs per year throughout project implementation, which were also supplemented by an indepth Mid-term Review (MTR) undertaken in 1993. SMs reviewed all aspects of the project, ranging from technical and institutional to financial aspects. Finally, the supervision process in the OTDP was efficient in identifying problems and issues.

However, the lesson learned from the supervision process of the OTDP is that in order to have a significant impact on project implementation and results, the co-operating institution needs to perceive its mandate beyond one in which it considers the supervision dimension from merely a mechanical point of view, to one in which it takes a pro-active role in support of project implementation. Supervision should not be delivered or received as an inspection function, but internalised as a project management tool available for improving the project's overall progress.

OTDP further illustrates that project supervision should increasingly not just focus on problem-identifying, but become a problem-solving exercise, where problems and solutions are identified in a participatory process involving the various project partners.

In this regard, experience sharing with project execution personnel and local experts as well as training through the organisation of workshop/seminars during or at the end of SMs with the participation of IFAD and other stakeholders proved to be useful.

For the supervision function to be more complete and efficient, it is proposed that co-operating institutions must also:

- Follow-up on recommendations: lay equal, if not more, emphasis in ensuring the implementation of recommendations made by the SMs. For example, the OTDP supervision reports highlighted the effects on implementation as a result of the withdrawal of the NGO (Agragamee) responsible for the Human Resources Development component. However, follow-up efforts on this issue (which was repeatedly reported in various SM reports) by IFAD and UNOPS in terms of identifying an appropriate alternative NGO was limited, which ultimately was a major factor adversely influencing the outcome of the project; and
- Composition: in multi-dimensional projects, like the OTDP, it is important to include specialists on SMs with varying expertise related to the project, as and when appropriate. For instance, with the exception of the MTR, and one other mission, OTDP supervision missions were composed of a financial analyst and an IFAD/UNOPS officer. The project would have benefited from the occasional participation in SMs of experts in agronomy, land rights, M&E, infrastructure engineering. Finally, co-operating institutions should take advantage in banking of local experts, not only to save costs but to ensure adequate input of local experience. This will facilitate the understanding of the local context and improve communication with project beneficiaries and local officers.

ROLE OF NGOS

There was one NGO (namely, Agragamee) initially involved in OTDP. It was responsible for the execution of the Human Resources Development (HRD) activities, a very critical component of the project, which included training and education of the targeted population in areas of prime concern related to the project. However, soon after the project become effective a power-struggle developed between the NGO and the project management team caused mainly due to differences of opinion on how project activities ought to be implemented. The NGO believed the peoples' voices were not given due emphasis in the prioritisation of projects' activities and implementation, whereas project management saw the NGO as misleading the target group by sensitising them in a (political) manner inappropriate for their participation in the project. This created an atmosphere of tensions and mistrust, leading after a few years of implementation, to the NGO's withdrawal from the project. Following this, a replacement NGO to implement the HRD component was not identified, and project management themselves conducted the earmarked activities, for which they were not well equipped. The evaluation mission believes that, the non-association for the full duration of the project, of an NGO with appropriate experience and capabilities to undertake HRD activities contributed to the shortcomings of the OTDP.

It may be noted that the NGO (Agragamee) involved in the project had a missionary zeal, a proven track record and experience of working closely with the target group, whose confidence the NGO was widely believed to enjoy. Another important point to note about the NGO is that, in addition to their attention to grassroots issues and concerns, their power base was also considerable amongst the tribal people. The latter factor appears to have led the NGO overtime to becoming somewhat unfocused in their thoughts, actions and operations.

While project designers do try to select the best NGOs in the project area, this selection must be made with utmost care, taking into account the NGOs' reputation and capability, but also their relationship with all the stakeholders. Furthermore, it is imperative to clearly define in advance the objectives, role and responsibility of the NGO and each partner in the process. To this end, it may even be advisable to identify during project design the NGO to involve, and to make them participate in the design process. In this way, it will be relatively easier to have a common understanding right from the start about what is expected of each partner. Finally, at start-up, special team building sessions could be organised to inject a spirit of cooperation and effective participation. This would also be an opportunity for all concerned to appreciate their respective roles and boundaries, as well as to recognise those of the other stakeholders. The project management and the governments on their side should avoid adopting paternalistic and top-down attitudes, and consider the NGO as an equal partner establishing a relationship based on consultation and dialogue. However, adequate monitoring mechanisms must be introduced at the project management level to ensure that the NGO involved is discharging its duties and role in a manner in which the overall objectives of the project under consideration will be satisfied.

Sustainability

Various activities were undertaken during the implementation of the OTDP resulting in the formation of economic infrastructure, such as roads, bridges, cross drainage works, community centres, water harvesting structures and small irrigation systems, as well as community plantations and the development of nurseries.

However, already at the time of the completion evaluation mission, the latter noted deterioration in the infrastructure work undertaken during the project period. Further, the future of the input supply activity is also under strain. In this respect, during implementation seeds of improved variety, agricultural implements and other inputs were provided free or at a highly subsidised cost. These are no longer available to the tribal people, and no mechanism was instituted to ensure the timely and regular supply of inputs once the project ended. The same applies to the soil conservation and agro-forestry sub-components, as the tribals were provided through the OTDP planting material and other equipment for these activities, which are no longer available. Finally, an important element having an impact on project sustainability is the issue of wage employment generated through the project during implementation, which ceased immediately after project closure.

In development programmes such as the OTDP, where there are a number of inter-linked activities such as infrastructure development, agriculture and natural resource management and human resources development, the crucial role of the beneficiaries, community-based organisations and counterparts cannot be over-stressed in ensuring the sustainability of the investments made. A lesson emerging from the OTDP is that without the adequate participation in the design, decision-making and implementation of the beneficiaries and their community organisations, the post-project sustainability can be of major concern. The degree of participation at all stages in the project was limited – for instance, tribal people were minimally involved in project design and were seldom involved in developing the annual work programme and budget. The project lacked consultation with the targeted population, thus creating an atmosphere of mistrust and discontent. Together with the issue of participation during design and implementation, OTDP also revealed the centrality of education, training and empowerment of beneficiaries and their organisations in meeting the objective of sustainability. This process leads to the much-needed sense of responsibility and ownership of the beneficiaries and their organisations in the project, and consequently willingness on their part to maintain project activities.

Further, at the design stage mechanisms should be introduced to promote (towards the end of the implementation period) the transfer of project activities to the target group, to increase the chances of sustainability. The role of the government in this context is of importance, not only in terms of their commitment, but also by providing for appropriate institutional support and an enabling policy framework. The involvement of the same institutions should also be considered in future projects to the extent possible, as this would also provide an opportunity for continuation of activities.

Equally important is the issue of financial sustainability. Governments need to allocate funds to sustain selected project activities after the closing date, and to this end they should introduce a budget line in their core programme budget for the purpose. However, to minimise reliance on counterpart funds for post-project activities, projects should be designed as far as possible in a way to ensure that most of the project follow-up activities would be undertaken without continued allocation of public resources to any significant extent.

Process Leading to the Finalisation

The following were the major steps involved in arriving at an understanding on the seven lessons learned from the completion evaluation exercise of the OTDP.

1. The first draft evaluation report was shared with IFAD's Asia & Pacific Regional Division (PI) on 24 June 1999, and ten proposed lessons learned were provided to PI on 28 June 1999.
2. Based on various discussions in-house, the report and the lessons learned were revised and shared once again with PI on 13 August, which agreed for the Office of Evaluation and Studies (OE) to discuss the report and lessons learned with the wider project partnership.
3. The draft report and lessons learned were subsequently sent to other stakeholders (Government of India, Government of Orissa, WFP, NGO and UNOPS) on 26 August 1999.

4. A brainstorming session was organised with various partners at the WFP premises in New Delhi on 13 September 1999 to discuss the draft report and proposed lessons learned. A consensus was reached on the most important learning elements emerging from the evaluation.
5. Additional written comments were received on 21 and 27 September from the Government of Orissa on the lessons learned and the report, which were used again to revise the documents.
6. The revised set of lessons learned were sent to PI on 12 October, and shared with OE evaluators for their comments and suggestions on 13 October.
7. The lessons learned were further discussed in detail with the Regional Division and others during the first Project Development Team (PDT) meeting on the design of the Second Orissa Tribal Development project at IFAD on 8 November. The PDT acknowledged "that the OE Orissa lessons also cater for more explicit and specific operational learning needs".
8. The final version of lessons learned were sent to the core partners on 18 November by email. Written statements were received (on 19 November and 22 November) from the core partners on a final understanding of the contents and nature of the lessons learned, and their commitment to utilise systematically the learning generated through the evaluation exercise when developing future interventions, policies and strategies aimed at tribal peoples development.

Understanding between the partners in the completion evaluation process, namely the State Government of Orissa (represented by the Scheduled Caste and Scheduled Tribe Development Department), the World Food Programme (as the Co-financier of the project) and IFAD (Asia & Pacific Division and the Office of Evaluation & Studies).

CULTURE AND SUSTAINABLE DEVELOPMENT

The search for truth and understanding in the nature of things; the attempt to "understand" the universe, questions about its structure, or, even more extremely, why does it exist is an argument which could continue indefinitely, with a series of generic questions and vaguer answers, without perhaps ever coming to any concrete conclusions, if we do not establish fixed points on which to hinge the reasoning.

For over two thousand years, these questions have created in man two different ethical positions; two ways of conceiving his own relationship with the cosmos. Atheists or agnostics on the one hand and men of faith on the other have opposed each other on these questions.

But the men of faith, the men who have welcomed the supreme being or, better still, God, in history, those people who have understood the revolutionary nature of this event, have often not known how to collocate correctly in their own lives material things and spiritual ones (give to Caesar that which is Caesar's and to God that which is God's). A knowledge and a culture, which is still partial, make man fearful and restless towards the contrast between the material being and the conscious being. The decay of matter and the incorruptibility of the spirit would seem to belong to a dipole which does not have a logic in common. Therefore could the Creator who planned the universe have given it two contradictory realities? One "bad" because it is corruptible and the cause of the difficulties which man has in life, and the other, in contrast, the only practicable reality? This is difficult to believe. This duality, to various degrees, has not only had repercussions of an ethical and moral order, but it has also been the cause of the various cultural events and customs of populations up until today and a fertile ground for the philosophers, thinkers and politicians who have addressed it, above all in these last two thousand years.

All of history, above all in the last two thousand years, is the story of the ethical-religious events of every civilisation, with various origins and traditions, but the substance has practically remained the same. Jumping from the traditions of those populations in which religion has found in asceticism the only means to lift the human spirit, to those where the advent of Christianity, though renewing and giving a new role and meaning to these forms, has given man the role of being present in history, in the world, in matter, to being near to it.

"In virtue of the creation and even more of the incarnation, nothing is profane down here for whoever is able to see". (Teilhard De Chardin P.) The evident social, cultural, scientific and, not least, political repercussions on western society are, without a doubt, influenced by these new and different methodological definitions.

However, a continuous and further evolution in culture in the third millennium can be had through a more complete and thus truer interpretation of the cultural and philosophical speculation which Christianity presents and possesses and which, though grafted onto history for around two thousand years, still has to make the plant of civilisation bear fruit in a complete and full way, transforming the raw lymph of human action into elaborate lymph.

In fact, "the work of the algae which concentrates in their tissue the substances scattered around them... is only a pallid image of the continuous elaboration undergone in us of all the potential of the universe, to transform it into spirit." (Teilhard De Chardin P.)

We know that the swaying of the pendulum is always due to a cause; the more energy that there is in the cause, the greater will be the effect; or better, in a certain sense, as in all aspects of culture and in the history of man, "man does not possess the faculty of instantaneously assimilating, of immediately digesting a novelty, a new knowledge, but in his progression (in his breathing) and perception of that new thing, he sways between two positions (as has often been the case in history) until, like the pendulum, he finds a point of balance (it is pedagogy in itself the fact that error and awareness of the sin leads to repentance, to the need to be included within the Great Principles)-"following the path which leads to the Mount of the Transfiguration." (Teilhard De Chardin P.)

Going back to what was said in the premise, what has matter got to do with all of this? And what relationship is there between it and man? Christianity leads man to intervene in the world, in places where there is need, through his contribution, in places where matter in all its possible and imaginable forms asks to be brought back towards the reason for its being.

Let's make this aspect clearer. We are often led to judge, to classify in airtight sectors, the things that we see: the tangible (we'll leave out the transcendental here), giving it various identities and qualities and often therefore different logic. We are spontaneously led to judge. For example, there are grasses which are beneficial for man and others which infest him; good and useful things and bad or even useless ones.

There exist parts of the world which are beautiful and good and others which are ugly and bad. This way of classifying the cosmos has its origin prevalently in a partial human valuation, but it is above all born from the presumption that this comprehension is sufficient enough to explain the world. This culture, which grew stronger above all in western countries after the years of the Enlightenment, thought itself able to explain through reason, and thus with the knowledge of the times, that which fell within the dominion of the senses.

It had in fact attributed an incomplete function to matter because made absolute and therefore out of every logic, of the logic which must necessarily give a role to matter as a counterpart to the spirit or even as the only reality in existence. This culture, which was useful in getting beyond the ideological waste of the Middle Ages (and thus in turn a transitional knowledge found between two eras) has led us to think in categories of things and not in wholes. This was a way of thinking which was useful for its time in the past, but transitory on the path of human history.

But if the tangible and transcendental universe is a whole, as it obeys the rules of a single planner, then for some time human logic, which is not different from divine logic (man is made in our image and likeness) has had to resolve a question (material-spiritual) without some necessary elements which are useful for the solution and for comprehension. It is as if in a mathematical equation we forced ourselves to find a solution without knowing the value of an unknown quantity, and it is the unknown quantity which gives a sense and value to the whole equation. (The unknown quantity is not however a synonym of inexistence as is the tendency in western societies). Let us now make another consideration.

What difference is there between a rock and a man? Let us argue this from an exclusively materialistic point of view. Unable to say the spirit (as it is an immaterial element), we would then be forced to talk about the ability to move (or immobility), to reproduce and so on. We would therefore be forced to make considerations about their substantial and visible nature and not on intellectual values. In effect, man (as solely a material component) and rock belong to the same universe, are made up of the same elementary particles, have to follow the same laws, the same rules, the same transience as matter. For too long Science, Technology, Culture and Politics have worked with elements and not with wholes, and the culture of elements behaves like an ostrich with its head buried in the sand. It does not see that the hole is contained in something (much more) bigger.

ETHICS AND SUSTAINABLE DEVELOPMENT

The word ethics correlated to the question of sustainable development becomes, in my opinion, the main principle through which to drive the whole ideological process that must be at the base of this great, epoch-making, yet to be implemented revolution. The global question of reference becomes trinomial: Economics, Ethics and Sustainable development, the interconnection of which is by now commonly recognised, represent three vast areas. As is natural, the specialists of each of these three sectors see this interconnection through their own particular point of view. Among all the point of views, the principal of the common good, and in this case, the universal common good, seems to me to be the most suitable to act as a link between the three elements:

economics, ethics and sustainable development. This principle requires that the global society be organised in such a way as to ensure that every man can have the possibility to achieve his full potential. Personal achievement depends on the effort made by all to look for the common good. In fact, the development we are speaking about-a sustainable development, considered as a component of an integral human development and which is based on three pillars, economic, social and environmental-must regard everyone, in the present and in the future.

In this universality, there is a double root: ethical and economic-functional. The ethical one is based on the principal of the eminent dignity of each human being, and for this it is opportune to direct the political principles towards the construction of a world in which each man, without exception of race, religion or nationality, can live a fully human life, free from the servitude which are imposed by humans and from a nature which has not yet been sufficiently controlled. The second root, the economic-functional one, sinks into the observation that, if development is not universal, if it does not reach every population, then it is not effective because it is deprived of the active contribution of many and because the areas of underdevelopment are, in the long term, a cause of unbalance which upset the positive dynamics of the development itself.

To achieve a development which has been conceived in this way, that is to say, both human and integral, one must never lose sight of the interior parameter of man, that parameter which is in the specific nature of the human being, a parameter which has been put aside by a materialistic culture; that corporal and spiritual nature which, in its duality, composes all of man. In this sense, the definition given by John Paul II is of interest when he raises the question of an "authentic human ecology", stressing how we worry too little about protecting moral conditions.

This is the reason why it is interesting to observe what is happening with regard to the question of sustainable development, the decisions and the actions which the international community takes and puts into practice in order to achieve it. In fact, since 1992, the year of the United Nations Conference on the "environment and development", better known as the Conference of Rio, the theme of sustainable development has been widely debated in the heart of the international community. It must be said that the start of reflections on this matter was promising, for the first principle of the Declaration of Rio states. "Human beings are at the centre of the preoccupations for sustainable development. They have the right to a healthy and productive life in harmony with nature".

Besides, if promoting the dignity of human beings means to promote rights-and in the question under examination, the right to development and to a healthy environment-this also means remembering the rights, that is, the responsibilities towards oneself, towards others, towards the gifts of nature which in any case is the place where human life is protected.

Now, in order to be sustainable, development must find a balance between the three objectives already mentioned: the economic, social and environmental ones, in order to assure the well-being of today without jeopardising that of future generations. Ecological sustainability is possible only in the context of social development and economic growth, and therefore the elimination, or eradication, of poverty, to use the terminology of international bodies, is a crucial component of sustainable development.

Yet, if it is true that poverty and squalor constitute a menace to sustainability in all its aspects, the contrary is also true. In fact, if today the greatest environmental problems are global problems, there can be no doubt that those who are hit hardest by this are the poorest populations rather than the well-off ones. Just to give a few examples: it is the poor who usually live in the worst environments, in the outskirts of cities or in shanty towns; it is again the poor who suffer the greatest damage in environmental accidents because they live in places which are more exposed to such accidents. Furthermore, many populations in poor countries acquire the essential resources necessary for living from agriculture. For them, the environment is not a luxury, a composite of essential means for subsistence: hunger, malnutrition, and forced migration derive from environmental degradation, such as the destruction of fishing or forest resources and so on.

For this reason one of the positive signs of our times is the permanent importance that the fight against poverty has assumed even within the international community. In particular, the ethical character of this struggle constitutes a meeting point for the international community. In the Declaration of the summit on social development, held in Copenhagen in 1995, three years after the Conference in Rio, the leaders of states and governments undertook at point 2 "to act in order to eliminate poverty in the world through national interventions performed with determination and through international cooperation, because we consider it as an imperative ethical, social, political and economical question for humanity". The situation in the world, however, especially among the poorest of the poor, is dramatic: it is enough to think that in 2000, in terms of human resources, 1.3 billion human beings were living below the poverty level, that is, with less than a dollar a day, while another

1.6 billion were living with less than two dollars. As is known, income is only one of the means of measuring poverty. The situation seems even more serious if we consider this phenomenon in a broader and more realistic sense, taking into account the deprivation of something; the lack of prospects in life, the lack of years of schooling, scarce health care, even basic health care, or the possibility to accede to drinkable water, not to mention more in general the impossibility to participate.

The international community understands this perfectly well, so much so that one of the first objectives of the so-called Millennium Objectives-indicated in a document signed by representatives of the UN, the OECD (organisation for economic cooperation and development), the International Monetary Fund and the World Bank-is that to halve the number of inhabitants of the planet living in absolute poverty in the period between 1990 and 2015. Since then, starting from the Millennium summit in New York in September 2000, there has not been a United Nations conference, or a conference of one of its specialized agencies, summits of heads of states and governments at a world or local level, among industrialised countries or countries in the process of development, which has not reasserted the priority of the fight against absolute poverty and the achievement of this aim.

In all this context, one of the other new phenomena, at least for its proportion, is globalisation which is neither good nor evil a priori, despite the fact that here and there exist stupid movements for or against it, even using the characteristics of the path of history in order to be involved in "politics" (even this is a sign of poverty). Some more evident characteristics of globalisation need, however, to be analysed, and among these is the increase in competitiveness, which produces a social harm that seems, at least for now, inevitable: the increase in inequality. In fact, the gap between the rich and the poor has become more evident, even in countries which are economically more advanced, and a sense of precariousness seems to be diffused, especially among younger generations.

In short, we are faced with a paradoxical situation in which, though the resources are not insufficient, as is globally recognised, thanks also to, we must admit, globalisation, the so-called relative poverty of three billion people has become more acute. Thus, apart from the case of countries which are very poor, the problem lies in an inadequate and unjust distribution of resources, for various reasons, at a national and international level.

For this reason, a "globalisation of solidarity" is necessary (cf the Vatican Document Centesimus Annus, 36) This ethical approach has to start with the question of the international debt of poor countries. But if governmental realism wants to recognise that the debt of some countries can not be collected-which in part is what has happened-it is important that the mechanisms that have been studied and already drawn up to give some solutions to both the creditor nations and the International monetary fund be applied at least within the time limits which have been established. It is also important to assure that the sums which correspond to the debt which has been cancelled are really employed by the governments of the debtor countries on social projects, above all on health and education.

One of the more durable ways to enact solidarity at a global level is that of bringing back equality in international commerce by eliminating protectionist barriers. More effort is needed to assure that every partner has the opportunity to gain benefits from the opening of their markets and from the unrestrained circulation of goods, services and capital. In fact, in today's world, commerce, development and the fight against poverty are closely tied.

Furthermore, it is universally recognised today that the key to development in general, and that of sustainable development in particular, is to be found in science and technology, and in these sectors the principal problems are the considerable obstacles in the transfer of "know-how" connected to technological progress from the rich countries, which have it to poor ones. If we consider that most of these are found in tropical areas where life expectancy is around fifty and if we remember that in the world more than 861 million adults, of which two thirds are women, are not able to read and write and more than 113 million children do not go to school, we can understand that those initiatives dealing with health and education are of utmost importance.

Though the negative effects of globalisation are mostly attributable to inadequate governance, also because of the inability to adapt at the same speed of the rapid changes in today's society, it is also true that some of the gaps in governance at a national level in poor countries are well known, above all to the inhabitants of those countries themselves. The first measures to take in order to fill them could be the following: resolve the numerous situations of conflict, the majority of which are ethnic in origin; reduce spending on arms; fight corruption and stop the flight of capital to foreign countries; encourage, as said, educational and sanitary programmes in order to create systems, even elementary ones, of social security. Above all in the field of sustainable development, it is also necessary to encourage the participation of the local population in their own development in respect of the subsidiary principle. In the poorest countries, some progress is being made in this respect, even if with great effort.

To give an example, the initiative of the Monetary Fund and of the World Bank to reduce the debt of poor countries which are highly in debt, known as the HIPC initiative and which has highly complex mechanisms, foresees, among other things, the presentation of an action plan known as the Poverty Reduction Strategic Plan (PRSP). It regards long term plans which must be elaborated by local governments with ample consultation with the civilian society.

It is needless to try to hide the difficulties that this consultation encounters, above all, as in many cases, when in the presence of governments which are not truly democratic and in countries where there are often no registers of births, marriages and deaths, where property rights are at the best of times uncertain and where it is difficult to understand what the land registries consist of. Despite this, it is positive to observe how the principle of participation has become a shared principle. At the level of global governance, the difficulties that a multi-lateral system which was created after the Second World War has in facing the complexities of a globalized world and the multiple hotspots of our days have never been more evident. It is enough to consider the strong protests seen at every G7/G8 meeting, the criticisms levelled at the international financial institutions or the composition and the working mechanism of the United Nations Security council. These criticisms are often the reflection of a positive consolidation in the sense of global citizenship, which becomes a reality in the increasing number and influence of Non-Governmental Organisations. Perhaps the time has come for these to play a more formal role in international public life.

In the field of sustainable development, faced with the deterioration of the environment and the fragmentation of the international institutions created by the many treaties on the matter, global governance has been called for by many sides. In fact we must recognise that though there exists a relevant organism in the United Nations, UNEP (United Nations Environment Programme), due to the mandate given it and for the scarce resources allocated to it, there exists at present a weakness which is quite evident in the so-called environmental pillar at an international level.

For example, it is necessary to have supervision of the enactment of multilateral agreements. One of the issues which becomes increasingly urgent in this sector is the scarcity of water, the fundamental element for human existence. It is a serious problem if we consider that if we proceed with the model of development used at present, around half of the world's population will suffer from lack of water in the next 25 years. These preoccupations emerged in all their seriousness during the Third World Forum on water held in Kyoto (16-23 March, 2002).

In all of this context can clearly be seen that the real nutshell of the question is tied to the formation of parameters for economic efficiency which the countries of the world give to their own economies; parameters which are still tied to a capitalistic system (which I would like to term first generation) which, up until today, has included in economic equations partial economic valuations without taking into account the economic valuation of the well-being of nature. The well-being of nature needs to be considered in the same way as the patrimonial inheritance which any person receives. Without an inventory and accounts of the assets, nobody can know how much they will receive in inheritance. The "first generation" capitalist system still behaves in this way, not giving any value (even monetary value if we wish, even if this is rather arid) to man and the ecosystem.

The ethical principle of sustainable development must include in its equation all those parts which, perhaps unknowingly, we have neglected until today. Just as today's system considers an economy that gives material richness to man as efficient, neglecting the most important part: a person's dignity.

6

Demographic Impact of Migration and Climate Change

The main purpose of this chapter is to review briefly existing research on the likely impact of environmental change on the movement of people. The topic of environmental migration is not a new topic of research. As early as 1990, the Intergovernmental Panel on Climate Change (IPCC, 1990: 20) warned that the greatest single impact of climate change could be on human migration – with millions of people displaced by shoreline erosion, coastal flooding and severe drought. In 1992 the International Organization for Migration (IOM) published a report on "Migration and Environment" in which it is stated: "Large numbers of people are moving as a result of environmental degradation that has increased dramatically in recent years. The number of such migrants could rise substantially as larger areas of the earth become uninhabitable as a result of climate change" (IOM, 1992).

However, until two or three years ago, the topic of migration and the environment was largely ignored by migration experts and policymakers. Indeed, in the 2005 report of the Global Commission on International Migration, there is barely a mention of the topic.

Part of this neglect may be due to the marginal consensus over the years among researchers about whether or not environmental migration is a distinct form of migration worthy of special study. There has been considerable disagreement about how to conceptualize the relationship between migration and climate change and about research methodologies to be used to investigate the topic further.

While it is recognized that there is a two-way relationship between migration and the environment, the main emphasis in this review is on research on the implications of environmental change for migration, rather than vice versa. This paper is based on the findings of a recent book published by IOM titled *Migration, Environment and Climate Change: Assessing the Evidence* (IOM 2009). Given limitations of space, the paper focuses on four main research challenges which are discussed in the IOM book:

1. How has the relationship between migration and the environment been conceptualized?
2. To what extent has it been possible to measure the scale of environmental migration?
3. What evidence is available regarding the impact of environmental migration?
4. What research has been conducted on policy responses?

The paper is not limited to a focus on climate change but looks more broadly at the range of environmental factors which impact on migration. This is because many changes in the environment, such as earthquakes, which are not necessarily linked to climate change, have an enormous impact on the movement of people.

CONCEPTUALIZING THE RELATIONSHIP BETWEEN MIGRATION AND THE ENVIRONMENT

The migration and environment research literature tends to fall into two broad and extreme categories:

1. work done by "minimalists" who suggest that the environment is only a contextual factor in migration decisions and
2. work done by "maximalists" who claim that the environment causes people to be forced to leave their homes (Fraser et. al., 2008).

Although many experts accept that climate change is a factor which can impact the decision to migrate, the conceptualization of this factor as a primary cause of forced displacement has been questioned (Black, 2001). While the environment can be a driver of migration, more often than not a complex combination of causes determines whether or not people move. Given the multiple causes of migration, drawing a clear line between voluntary and forced movements is not always straightforward.

This disagreement on the role of the environment in inducing migration is reflected in further disagreement over terminology. It is common to describe those who move for environmental reasons as climate change refugees or as environmentally displaced persons and to characterize such movements as forced migration. Popular with the media, the term "environmental refugees" has been used to describe the whole category of people who migrate because of environmental factors. This broad definition, while evoking an image that has brought public attention to the issue, is not sufficiently precise to describe all the various types of movements which may be linked to environmental factors. In some situations, such as natural disasters, people may have little choice but to move, and may be forcibly displaced. In other situations where environmental change is gradual, movement is more likely to be voluntary as people have time to weigh their options, and environmental change may be one of many factors inducing them to move.

It is perhaps more useful, instead, to think in terms of a continuum:

"Population mobility is probably best viewed as being arranged along a continuum ranging from totally voluntary migration… to totally forced migration, very few decisions are entirely forced or voluntary" (Hugo, 1996).

In the absence of an internationally agreed definition, IOM developed a working definition in 2007 which defines "environmental migration" as follows:

"Environmental migrants are persons or groups of persons who, for reasons of sudden or progressive change in the environment that adversely affects their lives or living conditions, are obliged to leave their habitual homes, or choose to do so, either temporarily or permanently, and who move either within their country or abroad."

The purpose of this definition is to try to encompass population movement or displacement, whether it be temporary or permanent, internal or cross border, and regardless of whether it is voluntary or forced, or due to sudden or gradual environmental change.

Measuring the Scale of Environmental Migration

How many people have been migrating in recent years due to environmental change and how many more people are likely to migrate in the future? There are no firm answers to these questions, but it is assumed that most of the migration that will occur will be mainly within those developing countries which are likely to be most affected by climate change. Today, approximately three-quarters of all migrants move within borders, and only 37 per cent of migration in the world is from developing to developed countries (UNDP, 2009).

Probably, the best available data on environmental migration are data on the numbers of persons displaced as a result of natural disasters. In 2008, for example, it has been calculated that 20 million people were displaced by sudden-onset climate-related extreme weather events (OCHA-IDMC, 2009). However, even in the case of natural disasters where better data exists, we have no global data on migratory movements related to natural disasters (Hugo, 2008). At best, there are estimates that can be derived from displacement data relating to particular crises (Naik, 2009). Despite the fact that the reported number of disasters has doubled over the last two decades (Basher, 2008) we have not seen a major impact on international migration flows, as much displacement is short distance and temporary. The Tsunami which hit Asia in 2004 claimed the lives of an estimated 200,000 persons and displaced around 400,000 others, yet the vast majority of those who were forced to move relocated to nearby areas.

Extreme environmental events such as cyclones, hurricanes, earthquakes, tsunamis and tornadoes tend to capture the media headlines, but gradual changes in the environment may have a much greater impact on the movement of people in the future. For example, during the period 1979 to 2008, 718 million people were affected by storms compared to 1.6 billion people affected by droughts (EM DAT, 2009). Unfortunately, however, there is relatively little information on the links between numbers affected by gradual changes in the environment and migration.

Reasons for Lack of Statistics

There are several factors which make it difficult to predict the likely scale of environmental migration (Brown, 2008). First, it is difficult, as mentioned earlier, to disaggregate the role of climate change from other economic, political and social factors which drive migration. Second, there is a basic lack of migration data available in developing countries which are likely to be most vulnerable to climate change. For example, in a recent report, the Commission on International Migration Data for Development Research and Policy noted that many countries still do not include basic questions about migration in their censuses. Even in the ongoing 2010 census, several countries including Japan, Mexico, Korea, the Philippines, and Egypt, do not include questions on the place of birth. One-third of countries also do not ask about previous residence in another country (CGD, 2009).

Third, the lack of data is largely due to the absence of an adequate definition to cover migrants affected by natural disasters under international law.

Fourth, it is extremely difficult to predict the impact of climate change. Climate modeling techniques have not yet even begun to account adequately for the impact of individual choice, the potential for international action and the variability of future emissions and meteorological scenarios (Brown, 2008).

Evidence of the Impact of Environmental Migration

There is relatively little reliable data on the impact of environmental migration because there has been relatively little empirical research on this topic. There are many good studies on the impact of migrants on environments, on land use, deforestation, and so forth, but there is almost no reliable evidence on the effects of environmental factors, "controlling for other influences, on out-migration, particularly from rural areas" (Bilsborrow, 2009). In a review of literature over the past 50 years, of 321 publications, including 153 articles in peer-reviewed journals and 29 books, only two articles were found which investigate the effects of environmental factors on out-migration based on quantitative multivariate methods (Moriniere, 2009).

Relatively few social scientists who focus on migration – and rely on data from censuses and household surveys – have been engaged in data collection or research on the environment (IOM, 2009). Moreover, the little research that has been conducted on the effects of environmental change on migration has tended to focus on the negative consequences of environmental migration. Few studies have explored how migration can be a coping or adaptation strategy or how migration can relieve pressure on environmentally degraded areas (IOM, 2009), as suggested in studies in countries such as El Salvador, Jamaica and the Philippines that have found that migrants respond to extreme environmental events by increasing their remittances (UNDP, 2009).

Research on Policy Responses

Research on policy responses to environmental migration is in its infancy. There has been little analysis of what "standards, policies or programmes are most appropriate for managing this category of internal or international migration flows" (Leighton, 2009). As most environmental migration is expected to occur within and between developing countries in the South, there has been little incentive for policymakers in destination countries in the North to adjust their immigration policies. Few destination countries have elaborated specific policy measures to respond to environmental migration, and none currently have a pro-active policy to resettle those affected by environmental disasters (Martin, 2009). At best, policies are ad hoc, with some countries taking measures to allow migrants to remain temporarily in the destination country when disasters occur at home.

A recent review of current policy responses in destination and origin countries finds that there is also little coherence between environmental change and migration policies (Martin, 2009). For example, few of the major middle-income developing countries which are major source countries for migrants, such as Mexico, India and China, have included any reference to migration in their climate change adaptation plans. Although many countries clearly lack adequate resources to respond to the growing number of natural disasters, the international community has at least established a policy framework for responding to emergencies. On the other hand, a strategy and policy framework to address the impact of gradual environmental change is largely lacking, and would require linking development, environment and migration policies in a much more coherent manner.

Analysis of policy responses to environmental migration has also tended to focus much more on responses to extreme environmental events rather than on how best to manage the impact of gradual changes in the environment on population mobility. Research tends to focus on questions such as how best to provide emergency assistance to those who are displaced, how to reduce disaster risk and how to improve the legal and normative framework for the protection of the displaced. A number of studies have also discussed whether there is a case for introducing a new set of legal instruments to protect the environmentally displaced. Zetter (2009), in a review of this discussion, concludes that there is little opportunity or need to create an entirely new set of legal instruments to address environmental migration. There is much scope within existing legal frameworks to provide protection to those who are forced to move for environmental reasons, but there is a critical lack of capacity in many states to implement existing frameworks and it is here where there is a need for much more research.

POPULATION, MIGRATION AND DEVELOPMENT

The relationship between MDGs and population trends and structures has started to be considered, but there is still need to improve our knowledge of how demographic mechanisms, including migration, influence progress towards MDGs attainment.

This book will mostly consider the situation in Pacific island countries and major migration countries in South, East and South-East Asia, including China and India. While the relationship between population, migration and MDGs is difficult to assess in large and interdependent Asian countries, linkages appear more clearly in small Pacific countries where the impact of migration is much more important. Although lessons learnt form the Pacific may not be replicable in larger economies, it contributes to improving our knowledge of the effects of migration on population and MDGs.

MIGRATION TYPES

Migration types are not neutral on its impact on population and development. In Asia, due to lack of political agreement on long-term and permanent migration schemes, such as green cards, migration consists mostly of 'guest workers' on short-term contracts that can be extended (Asis 2005, Abella 2005). Singapore has developed medium and long-term contracts for highly qualified migrants. Chain migration patterns have developed with unskilled construction workers migrating from Myanmar to Thailand; semi-skilled Thai workers migrating to Taiwan, Singapore, while Malaysians migrate to Singapore, Korea and Japan. Family reunification exists mostly for high qualified migrants. However, schooling of migrants' children has been addressed by several host countries including Singapore, Japan and Korea. Student and work contract migrations have consequences on age and sex of migrants. Students are youth; males migrate more often for industry jobs and women for services jobs. Age plays different roles according to types of migration as rim countries migration schemes grant points according to age. Despite forecasts of increasing shortages on labour markets in host countries (Mason), migration policies are still mostly oriented toward a restrictive selection process and temporary migration. This certainly reduces the level of legal flows and result in higher undocumented migration the characteristics of which are less precisely known.

In the Pacific, some Polynesian countries, except Tonga, are former New Zealand colonies and benefit from special access to New Zealand: Cook Islanders, Niueans, Tokelauans are all New Zealand citizens, and Samoans benefit from a special quota. Tuvaluans and Tongans (in Polynesia) and Kiribati (in Micronesia) also have some access to New Zealand under the Pacific Access Category (PAC), as does Vanuatu (in Melanesia)i (Bedford and al. 2006, Bedford 2005). Former US affiliated islands: FSM (Federated States of Micronesia) and RMI (Republic of the Marshall Islands) have special access to the US and to US territories (Guam and American Samoa). The special schemes have resulted in large scale migration, with free movements to New Zealand for selected groups of Polynesians, extended to Australia for New Zealand residents in the frame of the Trans-Tasman Travel Agreement, and resulting in the development of transnational communities with frequent circulation, as well as overstaying, between islands and New Zealand, further including Australia and the USii. Large scale circulation also applies to Micronesian migration towards the US. These two networks are typical of Pacific migration. Such schemes are made possible by the small size of island states and have resulted in mass migration from some island countries. But, although free movement would enable return migration and circulation, this has little developed beyond visits for Christmas and family events, and it has not resulted in much brain circulation that would enable more rapid economic development.

Labour migration has a long history in parts of the Pacific, initially for work on plantations, later for work in phosphate mines, and more recently through recruitments of seamen from Tuvalu, Kiribati and to a lesser extent, Fiji. There is also migration from Fiji to UK to work in military and to the Middle East to work as security personnel. Fiji is also largely affected by brain drain, with nurses migrating to rim countries and secondarily to Middle East. Shortages in nurses in Fiji have resulted in migration of Pilipino nurses. While Cook Islanders migrate to New Zealand, shortages in the tourism industry are filled by Fijians on short-term contracts. These are Pacific cases of chain migration.

DEMOGRAPHIC IMPACT OF MIGRATION

The impact of migration on population trends and structure is well known but flows have drawn most of the attention and, actually, a precise measure of flows is needed to assess the demographic impact of migration. Beyond uncertainties on the size of flows, data on stocks are also unavailable, with Singapore and Malaysia, among others, not publishing data on nonresident population. Would such data be available, comprehensive estimate of the impact of migration in a long-term perspective - answering questions like: what would be the population of the Philippines or the Cook Islands if there had been no migration since 1960 or 1970 - would be purely theoretical because fertility would certainly have been different on the long-term had migration not taken place. More simply, we shall try to consider the current impact of migration on population trends and structure from the latest censuses in Asia and the Pacific, with more attention to a few countries. Consequences of migration on population growth are well known. They include, as regards purely demographic impacts:

- Reduced population growth for emigration countries and higher growth for immigration countries - the former is usually considered favorable to developing countries;
- Changes in sex ratios according to gender differentials in migration patterns, with various directions being possible in both emigration and immigration countries - imbalances are usually considered unfavorable;

- Changes in age structures: increased dependency in emigration countries and reduced dependency in immigration countries – the former is considered unfavorable in the frame of the demographic window theory (Mason 2006, Mason 2001, Bloom, Canning 2001). Social and economic impacts are many, but the most commonly stated relate to:
- The labour force: depleted working ages and brain-drain from emigration countries and increased labour force and brain-gain in immigration countries, the former are considered unfavorable - however, migration releases tensions on the labour market;
- Family breakdowns reflected by increased headship rates for females in emigration countries and more frequent lone persons and 'not related' or 'other relatives' members of households in host countries – the direction of related effect is not well established

The impact of migration at regional level is more rarely mentioned but it deserves to be considered, mostly as it involves populations of very different size, like in the case of China and India on one side and South and South-East Asia on the other. Economies have become interdependent in the frame of a well established system of regional migration. In emigration countries with labour surplus, migration is a safety valve that reduces unemployment and underemployment, while in immigration countries with labour shortages, migration supports economic growth – however labour surpluses and shortages need to be considered at sectoral level.

The beneficial effect of emigration as a safety valve is limited in larger countries like China and India that can 'export' only a minor part of their unemployment and under-employment because receiving countries have not the capacity to absorb it and need to protect their labour market through migration and work policies (Rallu, 2001). For instance, in China, rural underemployment was estimated to be well above 100 millions people and the region, not even the world economies cannot integrate such numbers of unqualified migrants.

Nevertheless, the migration pressure of China in the region remains a concern for smaller South-East Asian countries, with Chinese setting up retail businesses and taking a large share of small trade. While there is a regional circulation of labour force at various levels of qualification and sectors of the labour market, there is also an extra-regional migration to Middle East or to developed countries of the Pacific rim and the West. Thailand, Myanmar, Malaysia, Indonesia, Bangladesh, Pakistan and Sri-Lanka are largely involved in intra-regional migration (as well as extra regional migration to the Middle-

East). But China, India, the Philippines and Viet-Nam have a large part of their migration directed towards rim countries and to the West.

At intra-Pacific level, migration of Pacific islanders is quite limited with only Fiji, Cook Islands and Palau attracting immigrants: mostly through regional organizations and the University of the South Pacific for Fiji and in tourism industry and construction in the Cook Islands (mostly Fijians) and Palau (mostly Micronesians). Fiji and Palau also have Asian migrants, mostly Filipinos, on work contracts in health and tourism. However, Cook Is and Fiji are mostly emigration countries. Palau experienced increasing emigration of its youth and Asian net migration declined in 2000-2005. Most of Pacific islanders' migration is directed towards rim countries: Australia, New- Zealand and the USA[iii] as well as to US and French territories but the latter rather consists in more or less closed US and French regional networks, that however include Asian migration, mostly for US territories. The mass migration from the Pacific, in the frame of preferential access to rim countries, results in important reduction of population growth in island countries.

Population Growth

The effect of migration on population growth varies greatly by countries, small countries being more affected than large countries.

In Asia, net migration remains below 2 per 1,000 and often below 1 per 1,000 in absolute values in most countries and therefore only slightly impacts on population growth. Its impact on China and India (-0.3 per 1,000 and -0.2 per 1,000) is even much lower (table 2). The highest negative net migration rate is observed in Sri Lanka reaching close to -0.5 per cent and the highest positive net migration is observed in Singapore reaching close to 1 per cent (it was close to 2 per cent for the 1995-2000 period) – Brunei Darussalam has migration rate of 2 per 1,000. In the Pacific[iv], net migration nearly erases the effect of natural growth in FSM, Nauru, Samoa, Tonga, or even inverses growth in Niue, Tokelau and occasionally in Cook Is[v] (Rallu, Ahlburg, forthcoming). In those countries net migration is frequently close or even above 2 per cent. Migration also considerably reduces population growth in RMI but it has declined recently. In Fiji, migration, mostly of Indo-Fijians but also increasingly of indigenous Fijians (with emigration rate of -1.7 per cent the former and -0.2 per cent for the latter; -1.0 percent for Fiji), reduces growth substantially and results in population decline for Indo-Fijians. Seamen migration from Kiribati had not much impact on growth as returns at end of contracts and some returns from Nauru tend to equilibrate flows. Palau had higher net migration rate in 1995-2000 (1.2 per cent) lifting total growth to 2.1 per cent, but it has much reduced in 2000-2005.

Sex Ratios

In the past, mostly males migrated and the analysis of sex ratios imbalances was used to reveal migration. But female participation in migration has been increasing for several decades and now many countries have higher female than male migration. Sex ratios have become difficult to interpret in regards of migration, mostly when emigration and immigration are present together. However, a few countries, like Indonesia, still have predominantly male migration and show low sex ratios at young adult ages. In the Pacific, ancient emigration countries (Tonga and Samoa) have turned to predominantly female youth migration, but mid-adult ages in Tonga show clearly imbalances in sex ratios linked with mostly male migrants in the early stage of the migration process. Predominantly male migration to Palau steeply increases sex ratios, reaching 139 males per 100 females at ages 25-44.

Age Structures and Dependency Ratios

The depletion of young adult ages linked with migration is a major demographic impact of migration. It is associated with change in dependency ratios. However, dependency is also strongly affected by the level of fertility. The impact of migration at young adult ages is sometimes difficult to see on agepyramids, due to past changes in fertility and various events affecting the history of countries, like conflict for Sri-Lanka. When migration affects a large range of ages like in the Philippines, the impact is not much visible. It is much clearer in Indonesia for males and Sri-Lanka for both sexes. The depletion of young adult ages is most typical in mass migration island countries where it results in bottle neck shaped age-pyramids, like in FSM, Cook Islands and Palau. In Palau, both youth emigration of both sexes and predominantly male immigration at mid and late adult ages are well visible.

Cohort Change

A more interesting approach to the impact of migration on population consists in cohort change. While age-pyramid and sex ratios did not show much effect of migration on population structure in the Philippines, changes in cohort size at ages under 50vi clearly show emigration in cohorts reaching ages 15 to 34. The impact was limited at ages 20-24 for males and 20-34 for females in 1995-2000, but it increased and extended to ages 30-34 in 2000-2005 for both sexes. In both periods, net losses are more important for females than for males. Some return migration appears at ages 35-39, but trends thereafter are difficult to interpret as combined effects of immigration/return migration and emigration can occurvii. Results may also be affected by quality of age reporting or age selective variations in coverage of enumeration.

Similar data for Pacific islands show much higher net losses starting as early as late teen ages, with one third of female youth cohorts and close to 40 per cent of males 25-29 emigrating from Samoa (and a similar situation occurs in Tonga and Cook Is), 20 per cent and 25 per cent of respectively female and male Marshallese aged 20-24, and between 15 per cent and 20 per cent of the 20-34 years old leaving Fiji, with higher rate at ages 30-34 in the frame of post-coup emigration that affected many adult Indians. In Palau, while all cohorts showed net migration increases between 1995 and 2000, only the cohorts reaching 25-29 in 2005 show net migration in 2000-2005, with emigration of Palauans at 20-24 and net departures of Asian migrants at ages above 30.

Such migration levels look like an exodus from small island countries and the situation in Sri Lanka and smaller Asian countries is intermediate between those of larger S-E Asian and small islands countries. At high levels of emigration, the impact on population structure and dependency is considerable.

Dependency Ratios

The increase in dependency due to migration is the result of depleted adult cohorts. Their impact on dependency can be high and it directly affects potential for development in the frame of the theory of the demographic window of opportunity. Recent studies have shown that low dependency is favorable to economic growth (Mason 2006, Mason 2001, Bloom, Canning 2001), on condition it is accompanied by high employment level of the youth bulge, which is possible in the frame of friendly economic and investment policies as well as social and political stability and good governance. But, emigration influences dependency ratios in the wrong direction. However, emigration countries have their workers abroad sending remittances that are usually higher than what they would earn in-country.

It would need controversial reconstruction of population trends for decades with and without migration to estimate the long-term impact of migration on dependency ratiosix. Therefore, the impact of migration cannot be estimated comprehensively and precisely, but it is possible to compare dependency ratios according to countries' migration status. In Asia, Singapore, China and Thailand have the lowest dependency and also the lowest fertility among countries listed in table 3. Among countries that have rather similar levels of migration like Bangladesh and Indonesia, the latter has much lower dependency due to completed fertility transition while the former has still TFR above 3. Similar characteristics and gaps prevail between the Philippines and Viet Nam, with the latter having both lower emigration and fertility resulting in lower dependency. And despite higher migration than the Philippines, Sri Lanka has lower dependency due to completed fertility transition. Further comparison and interpretation is difficult as several host countries (Malaysia, Singapore) do not include migrants in their census reports that relates to de-jure population only. Altogether, it appears that dependency is still strongly affected by fertility levels and the related children burden and the role of migration is difficult to measure, although its impact is evident.

In the Pacific, Palau is the only country that has already achieved fertility transition (with TFR of 1.9 in 2001-2005); it also has significant immigration and consequently shows the lowest dependency ratio in the region, similar to China and Thailand. In Fiji, Indians have achieved fertility transition but they have high emigration; their dependency ratio (55 in 1996) was much above that of Palau, but still much lower than for indigenous Fijians (70) who still have TFR of 3.3 and much lesser migration All other countries, except PNG, Solomon Is and Vanuatu are more or less affected by emigration. Samoa and Tonga have TFR just above 4 since the late 1970s and the mid 1980s respectively but they have ancient and important emigration. Their dependency ratio is similar to those of Solomon Is and Vanuatu that have witnessed fertility decline much more recently with TFR still close to 5 at the end of the 20th century. Despite similar, or even more pronounced, migration patterns than Tonga and Samoa, the Cook Islands have lower dependency due to lower fertility. FSM and RMI had both recent fertility declines and important migration; they also show dependency levels close to those of Melanesian countries.

Trends show that stabilization of the size of the population in the most affected migration countries in the Pacific goes hand in hand with nearly stable dependency ratios. Only RMI, FSM and to a lesser extent Fiji and Tokelaux show significant declines in dependency that are linked with fertility declines. Other countries show stable (Tonga), fluctuating (Cook Is, Niue) or slightly increasing (Samoa, Tuvalu) dependency ratios at a rather high levels, mostly when fertility remains high with TFR around 4. However, such stable fertility levels are necessary to avoid population decline. Altogether, these countries have achieved quasi stable populations through both quasi stable fertility and migration. As remittance based economies, achieving the demographic window does not seem to be felt as an issue in those countries. Thus, it appears that migration can erase the benefits of fertility decline for high emigration countries for a long period of time, eventually resulting in quasi stable population size and structure characterized by high dependency in the Pacific. Whenever high fertility is a major factor of high dependency in the Pacific, as well as in Asia, it appears that migration can delay the onset of the demographic window of opportunity, with most origin countries having dependency ratio well above 60 and as high as 80. The impact of migration on labour force and gender empowerment will be addressed in relation to MDGs. The above analysis of the impact of migration on population will help us understand its impact on MDGs attainment.

IMPACT OF MIGRATION ON MDGS

MDGs are development indicators and some of them have been recognized as such for decades, like infant mortality rate (IMR), as well as maternal mortality ratio (MMR) that appears to be closely correlated with the former in both Asia (Choe, Chen 2006) and the Pacific, and both are also correlated with the coverage of services in health. Other MDGs are obviously and closely linked to development, like poverty and environment (access to improved water and sanitation) indicators. Education is also a well acknowledged factor of development as well as of reduction of infant mortality and morbidity, mostly in relation to female education as well as employment. Thus, this section will address the impact of migration on development through its impact on population and MDGs indicators.

Services Indicators in Education, Health and Environment

The first demographic impact of emigration is reduced population growth and it is considered to have favorable effects. The impact of rapid growth on MDGs services indicators: primary and secondary education enrolment, immunization, skilled attended deliveries, is usually negative. In the Pacific, countries that have high population growth rates usually have lower coverage of services than those with low growth and they face the challenge of increasing coverage while births and children cohorts are increasing – not withstanding necessary improvement in the quality of services. This is clearly the situation in Solomon Is, Vanuatu and other rapid growth countries that have no emigration outlet. In migration countries like Samoa and Tonga, despite TFR still around 4, migration of reproductive age adults has resulted in stable birth cohort size since several decades. Thus, it has been possible to increase coverage of services close to 100 per cent and there is no difficulty keeping this level steady. However, this is realized through fragile balance between fertility and migration. Despite emigration, high and slowly declining fertility in the Philippines resulted in increasing birth cohort size until the 1995 census. The 2000 census is the first one to show stabilizing births cohorts at the basis of the agepyramid, with still the possibility of some under-enumeration of the

0-4 age groups, as shown in previous censuses. The same effect of population growth applies to environment services: access to improved water and sanitation. But it is complicated by urban/rural aspects. It can happen that urban growth consumes most of resources for infrastructures, leaving rural areas deprived, which is actually hindering rural development and increasing the urban drift. – The same probably applies to services in health and education, but water and sanitation indicators are usually disaggregated by urban/rural while it is more rarely so for education and health services which is a real gap in MDGs monitoring and may hide consequences of rapid urbanization.

Education, Child and Maternal Health Indicators

Whenever reduced growth can have favorable effects on education and health services indicators, brain drain depletes the stocks of qualified teachers and nurses. Transition rates from class 1 to 6, or up to form 3 – that some countries have included in their extended MDGs targets-, tend to be low and occur in the frame of many repetitions. A majority of students enter secondary education late by 1 or more years which jeopardize their chances of completing upper secondary education as well as technical education. Low qualification of the population results in difficulty for migrants to migrate through point systems, having to refer to short-term contracts or even migrate undocumented, and face unstable employment and low wages in host countries, which will affect remittances they can send home as well as their poverty status in host country.

IMR and MMR are dependent on coverage and quality of services and therefore they are affected by population growth. However, reducing child and maternal health indicators to very low levels needs high quality of services that is not available due to brain drain of qualified nurses and doctors. Therefore, IMR is stagnating between 10 per 1,000 and 20 per 1,000 in Polynesian migration countries, but it is above 30 per 1,000 in FSM and RMI. As Polynesian countries had already low IMR in the 1990s, achieving the MDGs targets implies rates below 10 per 1,000 that seem difficult to reach.

Gender Indicators

The second impact of migration on population identified above was imbalances in sex ratios. The effect of the sex distribution of migrants, with current increase in feminization of migration, on gender equity and women empowerment has been abundantly commented in the literature (UNFPA, SWOP 2006), but there is not yet a precise and comprehensive quantification of its effects on gender. For women, migration is sometimes a first chance of getting paid employment, but in case of return migration, they may not find similar job opportunities at home. Women left behind, despite remittances from migrant husbands, often need to work which increases the share of women in paid employment and is considered to be empowering them - female labour force participation has also favorable effect on health status of women and children, but it can be detrimental to children education. However, migrants' wives frequently live with relatives or in-law which limits their empowerment. A major benefit of migration for both males and females is behavior change that usually goes with living in a different economic and social environment which can have positive effect on gender equity and increase education of girls. However, this mostly works for the more educated. There is also evidence of migrants' wives living isolated in ghettos and experiencing sometimes even slower social change than in home countries. Social remittances or the dissemination of new ideas through visits of parents/ relatives to host countries and return migration is considered to be mostly missing in the Pacific, despite high mobility of islanders, and the proportion of females in parliamentary seats remains the lowest in the world. Altogether, like urbanization, migration is certainly an opportunity for women. However, this does not occur equally for everyone. There are social groups for which migration does not improve the situation of women and can even have negative effects. This is closely linked with education and economic status and calls to consider the impact of migration on MDG 1: poverty.

Poverty

The central factor in the relation between migration and poverty is remittances. However, labour force participation, a major determinant of remittances, disserves attention in origin as well as in host countries.

The underlying issue of the role of remittances on poverty is whether remittances reduce inequity of income distribution. The development of migration occurs through different phases with different impact on income distribution. At the beginning of migration, there is a 'migration hump': the more educated and wealthy have more opportunity to migrate in the frame of individual migration. In the first phase of migration, the proportion of households with one or more migrants in the lower income quintiles is lower than for higher quintiles and the amount remitted per household is also lower which contributes to increasing the inequality of income distribution. This is the situation shown by Fiji 2003 HIES data for indigenous Fijians whereas migrants and remittances are more evenly distributed by income quintiles for Indians with more ancient and higher migration (Brown 2006). However, this does not mean that there is no impact of remittances on poverty, as some remittances can actually be enough to lift a few households out of poverty, or anyway reduce the depth of poverty. At a later stage, lower social strata access more frequently migration and the impact of remittances on poverty reduction becomes larger. However, *in Tonga, although well advanced in migration process, only 33 per* cent of households in lowest quintile of value of non-land assets had a migrant against 55 per *cent to 75 per cent for higher quintiles* (Brown 2006, Ahlburg 1996). This is not taking into account effects such as increase in commodities, services, food prices and altogether the cost of living, or inflation, that may result from migration and remittances at the national level, pushing higher the poverty lines.

In Asia, the dominant pattern of low qualified labour migration departs somewhat from the phase pattern of migration and remittances on income distribution that prevails in individual migration patterns. Labour migration of unqualified or semi-qualified workers, may speed access to migration of lower (but not the lowest) income people. However, despite high levels of remittances nationally, amounts remitted per household for the poorest segment of the population is certainly low. Whenever, it may contribute to somewhat reducing poverty, it is unlikely to result in much economic development. Countries that rely mostly on unqualified migration show much slower economic growth than those where domestic economy is striving. The former are also more affected by brain drain as local wages are too low to deter qualified workers from leaving and attract return migrants. This is the case of the Philippines, Viet Nam, Indonesia, Bangladesh, Pakistan and Sri Lanka that have much lower GDP per capita than Malaysia and Thailand. Despite low GDP per capita, China and India attract return migrants in the frame of very large differentials in sectoral markets, with high profits being possible in the new technology and export sectors. Pacific island countries have sluggish economic growth and high cost of living which, associated with low wages does not make return migration attractive for nurses and teachers.

Although developed countries are usually not considered in the frame of MDGs (except for Goal 8) as host countries of international migrants, they are concerned by poverty issues as well as population living in slums (MDG 7). Unqualified migrants have often the lowest jobs (3 Ds) and wages in host countries and sending remittances push many migrants' households into poverty. Data from US censuses show that economic position of Pacific migrants, primarily from Samoa, Tonga, Micronesia and Fiji, improved in the 1980s, despite unfavorable macroeconomic conditions, and improved further in the long economic upswing of the 1990s. Part of the improvement in the position of Pacific migrants was due to increases in their human capital: education, work experience and language skills. These gains allowed more workers to acquire white-collar jobs and to increase their earnings (Ahlburg, Song 2006, Ahlburg 2000). Thus, the poverty rate of Pacific Islanders fell from 1.57 times the US rate in 1990 to 1.33 times the US rate in 2000. As regards remittances as well as for the living standard of migrants, migration is obviously more profitable when it consists of qualified workers. Thus, there is a kind of vicious circle with brain drain affecting the production of qualified migrants. Like for other workers, the socio-economic situation of migrants is deeply affected by their education status.

Employment Indicators

MDG indicator 'employment-to-population ratio at ages 15-64' under goal 1 has replaced 'youth unemployment rate' under goal 8. The impact of migration on the old and the new indicator is not much different although it is stronger on the former because migrants are often youth, mostly in the Pacific. As data for the new indicator are not yet available for all countries, we shall consider the old indicator. Migration reduces tensions on the labour market. Whenever, migrants may be employed before migration, their departures make jobs available for non migrants, as long as the economy is not too much affected by out-migration. Moreover, many young migrants never worked before migration, which is typically the case in the Pacific in the frame of preferential migration to New Zealand. However, youth unemployment is still high in these countries, from 12 per cent in Samoa, 30 per cent in Tonga, 35 per cent in FSM and 63 per cent in RMI. This shows that migration without economic growth is not the solution to unemployment. It is still more the case in Asia where emigration countries have actually the highest youth unemployment rates (around or above 20 per cent in Indonesia, Sri Lanka and the Philippines, and migration can only have a marginal effect on unemployment in China and India due to the size of their populations. Economic growth, not emigration, is the best way to reduce pressure on labour markets. This may be a reason why China and India have developed circulation of elites and return migration which contributes to economic growth.

Health and Environment

In origin countries, beside brain drain, migration also reduces the size of labour force in the lowest wage sectors, among which agriculture, reducing agricultural production and implying food imports. In the Pacific, migration and remittances have caused unprecedented extent of changes from the 1960s: modernization of housing, changes in diet and live styles that have resulted in the last decade in epidemic levels of NCDs: cardio-vascular diseases, cancers and diabetes. Dual mortality patterns have appeared in most of Pacific island countries as communicable diseases are still frequent due to poor quality of health services. The high cost of treatment of NCDs uses increasing proportions of the health budget hindering improvement in primary health care. At the end of the day, the impact of migration and remittances on population health as well as on the health budget is probably negative in the mass migration countries of the Pacific. This has led these countries to include NCDs in MDG 6 beside HIV/AIDS, TB and malaria.

Migration and remittances also impact on environment (MDG 7). Changes in consumption patterns result in more toxic waste and pollution affecting air, fresh water resources and coastal reefs. In Polynesia, shortages in agricultural labour force and remittance money have led to inconsiderate use of pesticides that affects coral reefs and reduces sustainability of in-shore fish stocks. Although poverty is high in RMI and FSM, the increase in automobiles and other household equipments have resulted in large amount of waste that cannot be dealt with in atoll environment and causes chemical pollution that can affect fresh water resources. Lack of adequate sanitation and domestic piggeries in Kiribati have led to high bacteriological pollution of the lagoon in South Tarawa, while pollution from automotives is also increasing. Population density in Ebeye (RMI) reaches above 30,000 per sq.km, it is above 10,000 per sq.km in Majuro urban area and above 8,000 per sq.km in Betio (South Tarawa). These environment issues are also present around cities in other Pacific islands and in Asia.

The relation of HIV/AIDS and TB with migration has already been largely debated, mostly as regards the former. It is true that migration increases risky behavior among lone migrants and results in HIV infection that is transmitted to spouses and partners. Behavior change, women empowerment, safe sex, anti-drug information campaigns are needed to reduce and revert the spread of HIV/AIDS. There is also urgent need to develop relevant information to eliminate inaccurate ideas and discriminative practices against PLHA (people living with HIV/AIDS) as well as migrants altogether. Furthermore, various MDGs indicators can be improved through the many uses of remittances:

- Purchase or in-kinds remittances of telephones, cell phones, computers and internet connection - Pacific countries with the largest expatriate communities are also those with the highest use of internet;

- Payment of education fees for children left behind or other relatives;
- Payment of health expenditures for children, parents or other relatives;
- Use of remittances for collective purposes:
  - Infrastructures (wells, improved toilets, sewage, generator/solar power, school books/materials, medicine for health centres), disaster relief,
  - Social life: cultural, sports or youth associations, churches..., that can contribute to social change and women empowerment and improve gender equity.

Finally, remittances used for investment or savings contribute to employment generation and economic growth in various ways, with related impact on MDGs indicators. Altogether, whenever migration reduces, if not erases completely, population growth easing increases in services coverage, changes in age structure and dependency due to migration have adverse effects. It delays the advancement in the demographic window of opportunity that is favorable to development. Sectoral shortages in labour force can appear, mostly nurses and teachers for which international demand is high. Brain drain hinders improvement in quality of services causing stagnating infant and maternal mortality and low levels of education and qualification of the population and future migrants, creating a kind of reproduction of marginality.

EMERGING ISSUES

In the frame of the space available in this paper, we shall briefly consider two emerging issues: one relating to bride migration and the other to brain circulation, social remittances, governance and development. Although they are rather already old issues, there has not been much progress done in these areas.

Bride Migration

Bride migration is already an issue in Asia. It is not unknown in the Pacific, mostly in countries where families used to arrange marriages and this now occurs between islands and rim countries. Imbalanced sex ratios at birth have increased in Asia. Female deficit on marriage markets is already felt in China and India, as well as in Korea (Rallu 2006). Given the size of the deficit in China (above 1 million from 2015, or more than a female birth cohort in the Philippines or Viet Nam) and India, it is unlikely that bride migration from neighboring countries can be the solution. However, even limited, bride migration would severely deplete female marriageable cohorts in other countries of the region. This is only one side of the problem. The other side would consist in continued imbalanced sex ratios at birth in these countries based on the belief and hopes that migration will fill the gaps. A continuation of the trends, or stabilization of sex ratios at current levels, would have an impact on population growth and ageing (Attane 2006, Cai, Lavely 2003) as well as consequences on family, society and culture.

Brain Circulation and Social Remittances

The guest worker and highly qualified migration patterns in Asia have moderately developed into brain circulation. Brain drain and sectoral shortages are still increasing. China and India have succeeded in developing brain circulation at a high level of qualification, with PhD students and young researchers studying and working in the US in ICT (often on Chinese or Indian data for social sciences) and returning to teach or work for a few months in universities in their countries of origin. Such pattern is not much developing in other countries for two types of reasons: lack of interest of migrants in returning to work in slowly developing economies, even for short periods, and lack of policy environment that enables such exchange, in the frame of suspicion of administrations, specifically in the Pacific, towards people who have been in contact with other ideas and could try to change the traditional, hierarchical and political systems that prevail in origin countries.

The lack of brain circulation is only part of a larger context that consists in the lack of social remittances in the Pacific. Despite large transnational communities with frequent movements between islands and rim countries, migration has not resulted in much social change in island countries and even only limited change occurs in migrant communities in rim countries. While some migrants leave to escape the constraints of local societies as regards traditional and religious life, most migrants organize their communities around ethnic lines and church leaders in rim countries. Return migrants who set up businesses and do not want to comply with traditional and religious authorities and the custom of gifts to the chiefs, are subjected to various pressures and discriminations and often re-emigrate. Altogether, island countries benefit from remittances, but do not reap the benefits of social and political change that would make business environment more attractive. Migration acts rather well as a safety valve for social problems like unemployment and lack of economic growth so that the need for policy change is not felt, or even is felt as a threat to the traditional system of governance. The low proportion of females in parliamentary seats in the Pacific, actually the lowest in the world, is testimony to the lack of social remittances and related limited women empowerment that does not reach to highest decision making positions.

CLIMATE CHANGE AND MIGRATION

Climate experts assume that global warming may well lead to a global shortage or at least to a displacement of cultivable land. The reasons for this are as diverse as are the impacts of climate change; rising sea levels will lead to more frequent flooding and storms in coastal and delta regions, small island states and low-lying coastal regions could disappear completely as a result. In some regions rainfall will increase significantly, provoking periodic flooding, while in other regions precipitation will quickly decrease, rapidly promoting droughts and desertification. Soil erosion caused by sandstorms and the decline in vegetation will reduce agricultural productivity in these areas — often already low — to a minimum, potentially endangering the food supply for major regions as a whole. The creeping effects of global warming and the associated increase in extreme weather events, as well as the deterioration in living conditions, may give rise to new migration streams.

This policy brief deals with the phenomenon of *environmental migration*. It focuses entirely on the effect of climate change on global migratory movements without neglecting the environmental consequences on the regions of origin and destination. The following paragraphs will firstly contain a comparison of estimates as to how many people will be affected worldwide and an introduction of those areas where climate change is most likely to cause migration. The brief will then examine the two main controversies concerning this phenomenon: the causality relationship between environmental factors and new migratory movements as well as the legal position of the persons concerned. The conclusion emphasises the necessity of extending the protection of people affected by the phenomena of climate change at the international level, even if it cannot be assumed that there is an exclusive causal relationship between climate change and migration.

Estimates

Reliable statistical data cannot be collected as there is not an internationally recognised definition for the phenomenon of climate-induced migration. In addition, estimates are also hindered by the fact that an immediate connection between the consequences of climate change and migration cannot be clearly demonstrated. In the absence of authoritative forecasts, there is a series of estimates based on unsupported assumptions (so-called *guesstimates*). The figures vary depending on which climatic, demographic and social values the estimates are based on. Under favourable conditions, there may be only a slight increase in current migratory movements, but under unfavourable conditions even high estimates appear to be too low.

In 2002 the UNHCR estimated the number of people forced into migration as a result of flooding, famine and other environmental factors at 24 million and later the number of persons displaced internally as a result of natural catastrophes alone at 25 million. The German Advisory Council on Global Change (WBGU) assumes that 10-25 % of all global migratory movements are the result of climate change and its consequences; that would be the equivalent today of an absolute number of 25-60 million migrants. *The United Nations University – Institute for Environment and Human Security*, or UNU-EHS, in Bonn estimated the number of environmental migrants up to 2010 to be at least 50 million. The Intergovernmental Panel on Climate Change anticipates a total of up to 150 million migrants as a result of climate change by 2050. The United Kingdom's Stern Review bases its estimate on a review of a large number of studies and forecasts and concludes that there are likely to be 200 million environmental migrants by 2050. The figures of Oxford professor Norman Myers are also widespread; he anticipates more than 200 million environmental migrants by 2050.

Affected Areas

In addition to the estimates given above, even the size of the population in areas that will be particularly affected by climate change can provide a useful reference as to the number of people who will be facing special climatic challenges in future and who may possibly regard migration as an alternative. The United Nations standing committee responsible for determining internationally recognised terminology (*Inter-Agency Standing Committee*, IASC) has identified four important scenarios that are likely to trigger migratory movements:

It is a decisive aspect in all scenarios that climate-related migratory movements may take place both within the affected nation states and across international borders, and may be further assigned case-by-case to a continuum of voluntary migration, preventative migration and refugeeism. Such migration may also be either temporary or permanent. Endangered states are deemed in general to be the poorly developed island states (Small Island Developing States, or SIDS), the sub-Saharan states, Asian coastal states, the Polar region, African developing states (Less Developed Countries, or LDC), the least developed countries worldwide (Least Developed Countries, or LLDC), the Near and Middle East, and Central Asia. Depending on the nature of the consequences of climate change, areas affected in line with the IASC scenarios may be divided into the following categories.

AREAS AFFECTED BY SIGNIFICANT, PERMANENT LOSSES IN STATE TERRITORY

This phenomenon comes as a result of rising sea levels and will most probably affect the South Pacific island states in particular (Carteret Islands, Kiribati, the Maldives, the Marshall Islands, Palau, the Solomon Islands, Tokelau, Tuvalu and Vanuatu), which have come to be known as "Sinking Islands", but also low-lying coastal regions in Alaska and the Bay of Bengal. As a result of land losses and the salinisation of coastal regions, some states have already started to permanently relocate inhabitants of their island states, while other countries are not ruling out the possibility of the permanent relocation of all or large parts of their populations. The possibility of relocation to a receiving country or else the founding of new states on uninhabited islands or ceded territories could be considered.

Flood Areas

The rise in sea levels in particular, as well as its hydro-meteorological consequences (increase in periodic floods, tropical storms, coastal erosion, salinisation of coastal waters), represents an important possible inducement for mass-migration. This would affect coastal regions. In addition to small island nations,. According to the Stern Review, by 2080 between 10 and 300 million people will have been affected by the rise in sea level alone, assuming a temperature rise of between 2°C and 4°C.

The IOM estimates that an one-metre rise in sea level would affect 360 000 kilometres of coastline worldwide. Roughly two thirds of the world's population live no further than 100 km from the coast, and areas that lie a maximum of ten metres above sea level alone, the so-called *Low Elevation Coastal Zone* (LECZ), are home to 634 million people – nearly a tenth of the world's current population. Of these, 360 million live in large towns near the coast (in other words, 13% of the global population living in towns). Most of the people in the zone that is affected by rising sea levels live in Asia, Africa and Europe. A current study on the rate of urbanisation in the LECZ recently showed that, alongside the small island states, the densely settled and heavily urbanised deltas and coastal areas in Asia and Africa are particularly exposed to an increased risk of flooding.

Not everyone in the LECZ will have to leave their homes, but rising sea levels could place those in low-lying areas and areas near the coast in acute danger. According to a study carried out by the Potsdam Institute for Climate Impact Research, there are already about 200 million people living in coastal areas that lie less than a metre above sea level.

Thirty of the world's 50 biggest cities lie directly on a seacoast. In the event of a rise of just one metre, according to the study, Egypt's Nile Delta and close to a fifth of Bangladesh (with 35 million inhabitants) would be especially affected, as too would large areas of Suriname, Guyana, French Guiana, the Bahamas, Benin, Mauritania, Tunisia, the United Arab Emirates, Pakistan, India, Vietnam and China. In Europe, an estimated 13 million people would be threatened by a one-metre rise in sea level (especially in the Netherlands and Denmark), including about 3.2 million in the German flood plains. Should sea levels rise by up to one metre, as anticipated, people living in low-lying coastal areas and sea deltas around the world will have hardly any other alternative than to emigrate to other areas.

Drought Zones

Numerous other areas will in future have to contend with a shortage of drinking water due to climate change. The authors of several UN Millennium Ecosystem Assessment studies established that droughts, desertification and the associated decline in agricultural yields are among the strongest factors that will cause people from arid areas to migrate to other regions. The reason for this lies in the far-reaching impact of water shortage, which will bring with it difficulties in supplying drinking water, loss of harvest and health and hygiene problems.

Already today there are more than 1.2 billion people living in regions where there is a shortage of fresh water, i.e. where natural fresh water resources are insufficient to cover the needs of the people living there. This especially affects the northern and sub-Saharan states of Africa, the Near and Middle East, the former constituent republics of the Soviet Union in Central Asia, as well as South East Asia and extensive parts of North China. Some countries in Central and South America also already have to contend with a shortage of water. In all these regions the impact of climate change may lead to longer drought periods, desertification and substantial soil erosion.

REGIONS VULNERABLE TO CONFLICT OVER NATURAL RESOURCES

In addition to emigration movements, the impact of climate change may also lead to conflict over resources. An external WBGU report concludes that, where possible climate-related conflict is concerned, the core regions are in Africa, Asia and Latin America. The climate-induced decrease in cultivable land and water resources affects a population with a growing percentage of youth who already today are likely to migrate into the cities.

This could promote religious, ethnic and civil conflict. The number of inhabitants in regions directly at risk of conflict over resources is approaching one hundred million. If we then add to these the number of inhabitants in areas at indirect risk, the number of potentially concerned persons rises to over one billion.

No matter which trigger for possible environmental migration we examine more closely, those most severely affected will be the small island states as well as the LDCs and LLDCs of Africa and Asia.

But not all of the people living there will migrate for environmental reasons. Infrastructure measures to shore up the coasts, water management plans and new technologies might suffice in a large number of countries and regions to lessen the impact of climate change. Yet, even if only a few percent of the people affected by climate change become environmental migrants, their numbers may reach the scale of the currently estimated refugees and internally displaced people (IDPs) (as at the end of 2008: approx. 42 million).

THE CONNECTION BETWEEN CLIMATE CHANGE AND MIGRATION

The fact that the rise in sea levels or salinisation of coastal areas as *climatic processes*, or hydro-meteorological natural catastrophes as *climatic events*, may trigger migratory movements is not disputed. However, environmental migration does not result froma single cause, but rather incorporates complex interactions of existing social, demographic and political contexts. When considering migratory movements in association with climatic processes or events, therefore, a distinction must be made between climatic and non-climatic migration factors, since migration is not necessarily going to occur for reasons of climatic events alone.

In this regard, adaptation strategies play a decisive role, for a society's vulnerability always results from its particular risk situation in a geographic sense and the efforts such a society makes to adapt. Thus hydro-meteorological catastrophes such as floods or tropical storms only lead to relevant migration phenomena if there have previously been political and social failures to adapt to the specific geographical risk.

In the absence of early warning systems, cross-institutional rescue plans, flood plains or dams, a society's vulnerability in the event of hydro-meteorological catastrophes is increased, as evidenced by the impact of the 2004 seaquake in the Indian Ocean. The tidal waves of the resultant tsunami destroyed entire coastal regions in the Bay of Bengal and South East Asia. At least 165,000 people were swept to their deaths and 1.7 million were left homeless. Some of the main reasons for the devastating impact of the tsunami were the lack of an international early warning and information system as well as the uncoordinated and partially non-existent evacuation of coasts in the affected region. The razing of mangrove forests and elimination of flood zones in coastal areas, as well as their settlement, also contributed to the enormous casualty figures. Not only catastrophes lead to emigration. It is even estimated that the steady degradation of habitable land due to climate change will in future be the most important trigger for international migration. These predictably long-term consequences of climate change already represent a special challenge to the societies that may be affected, for the ecologically induced loss of habitable land is fundamentally "a social problem that can be avoided."

Environmental migration is related to issues that make migration not only necessary, but also attractive, the so-called *pull* factors. These may be of a demographic, social, political or cultural nature. Population pressure, poverty, poor social welfare systems as well as poor governance in states affected by climate change are as decisive triggers for migration as climatic conditions.

At the same time, environmental migration takes place in developing countries in an environment of urbanisation for economic reasons, making it difficult to distinguish environmental migration from "normal" migration in metropolitan catchment areas. Climate change is only one factor in a bundle of factors of varying strength. Migration itself can be interpreted as a means of adapting to the socio-economic and political realities under the conditions of a changing environment. In cases of particularly drastic governmental mismanagement this can mean that a climatic event serves solely as an inducement to migrate, although the main causes are of a political and socio-structural nature.

Environmental migration is therefore not solely based on a simple matter of cause and effect wherein migration is always triggered by climatic conditions alone. It is in fact much more complex than that. If we wish to understand the motives for migratory movement, then previously-existing *pull* factors in particular play a decisive role.

This mutual influence and overlapping of environmental factors with political, social and cultural aspects of migration means that it is not possible to differentiate clearly between voluntary and forced migration, which in turn affects the definition and treatment of people affected by environmental migration.

Categorisation of Affected Persons

There have been numerous attempts to find terminology and definitions for the migration scenarios described above. In addition to the term *environmental migration* used here, there are such expressions as *climate change migration, forced migration* and *environmental refugeeism*. In the English-speaking world the composite term *climigration* is increasingly common. As environmental migration also concerns a mingling of economic and ecological factors and it is virtually impossible to make a clear distinction between these aspects, some authors also refer to *ecomigration*.

The affected people are mostly referred to as *environmental migrants*, but also as *forced climate migrants, environmental refugees* or *environmentally displaced persons*. The terms used for affected people is of decisive importance for categorisation as a migrant or refugee and the resulting consequences with regard to the international obligation to protect or provide for such people. In contrast to migrants, refugees are granted rights by the Geneva Convention concerning aid and services of the United Nations High Commissioner for Refugees (UNHCR) and may not be deported by receiving states (non-refoulement).

The term *environmental migrant*, coined by the IOM, is finding increasing international acceptance. To facilitate an initial basis for further research and data collection on the phenomenon, the IOM presented a working definition, according to which environmental migrants are "*persons or groups of persons, who, for compelling reasons of sudden or progressive changes in the environment that adversely affect their lives or living conditions, are obliged to leave their habitual homes, or choose to do so, either temporarily or permanently, and who move either within their country or abroad*". This definition seizes on the dimensions considered by the IASC of duration, direction and voluntariness of the migration.

Scientists involved in the European research project EACH-FOR (*Environmental Change and Forced Migration Scenarios*) based their studies on a three-part working definition. They distinguish between *environmentally motivated migrants, environmentally forced migrants* and *environmental refugees*. The environmentally motivated migrants differ from the latter two insofar as their change of location is voluntary. The difference between environmentally forced migrants and environmental refugees lies in the fact that forced migrants are subjected to a planned and long-foreseeable, but inevitable migration, whereas climate refugees are forced into sudden emergency migration by catastrophic scenarios. The EACH-FOR working definition does not consider whether in addition to the consequences of climate change there are also social, economic or political inducements to migration, whether the migration is temporary or permanent or whether the migration is only internal or also includes crossing state borders. Like the IOM, the EACH-FOR study picks up on the idea of three levels of duration, direction and voluntariness, but emphasises more strongly than the IOM the possibility of there being mixed causes for migration.

Analogous to the term *Internally Displaced Persons* (IDP), the Norwegian Refugee Council pleads for the descriptive term *Environmentally Displaced Persons* (EDP). This description includes all persons *"who are displaced within their own country of habitual residence or who have crossed an international border and for whom environmental degradation, deterioration or destruction is a major cause of their displacement, although not necessarily the sole one"*. The NRC picks up solely on the aspect of direction, i.e. both internally displaced persons and international refugees are included in the definition. The organisation does not consider either the possibility of voluntary migration, such as is allowed for in the IOM definition. The variation of migration triggers are not relevant for the categorisation as a climate migrant, but only the fact that the consequences of climate change are the main trigger of migration.

Controversy has developed in expert circles in particular with regard to the term *environmental refugee*. The reason for this lies in the special legal protection enjoyed by refugees in accordance with the Geneva Refugee Convention (GRC) and additional protocols.

Essentially the question is whether persons affected by climate change should in future be granted refugee protection in accordance with the GRC and its additional protocols. Article 1 A(2) of the convention states that the term refugee shall apply to any person who *"owing to well-founded fear of being persecuted for reasons of race, religion, nationality, membership of a particular social group or political opinion, is outside the country of his nationality and is unable or, owing to such fear, is unwilling to avail himself of the protection of that country; or who, not having a nationality and being outside the country of his former habitual residence as a result of such events, is unable or, owing to such fear, is unwilling to return to it."* As soon as these facts have been proven, the person concerned is granted refugee status.

The UNHCR rejects the use of the terms climate and environmental refugee as a matter of principle, since it fears that the term refugee established by the GRC and its additional protocols could be undermined by the category *environmental refugee*. Other UN organisations that come together under the aegis of the IASC, as well as the IOM, fear that the introduction of the term *environmental refugee* may undermine the established legal instruments for protecting refugees.

The basic conditions for refugee status formulated in the GRC, i.e. the fact of persecution and cross-border migration, would not be met in the case of environmental migration. The impact of climate change does not as yet count as persecution, the majority of the affected persons are internal migrants and therefore still within the protection of their own country. They are therefore less in need of international aid than Convention refugees, according to the UNHCR.

The UNHCR points out that under some circumstances some persons affected by climate-induced migration would meet the conditions for the granting of refugee status in accordance with the GRC. If persecution can be proved for persons fleeing conflict caused by climate problems, then the refugee condition is satisfied. Citizens of the "sinking islands" could also satisfy the GRC conditions if they migrate across borders, because such cases would potentially be a new form of statelessness. If countries of origin were to lose their entire territory, the affected persons could then be treated as stateless and thereby fall under the protection of the Geneva Refugee Convention (GRC) and the attached protocols.

However, the granting of refugee status in the case of the sinking islands scenarios is disputed because it is closely associated with organised or intentional migration. Such intended or tolerated migration can be the result of governmental projects such as the construction of dams or the establishment of flood plains. Both voluntary internal migration (motivated by compensation payments) and forced relocation both within national borders and across international borders occur here.

Essentially, however, the UN Refugee Agency seems to be concerned with preventing the extension of its own mandate due to its already considerable burden at a time when it is financially stretched. It may indeed be one of the organisation's obligations, according to a UNHCR paper, to point out to the international community the gaps in the protection offered to the people concerned, but it is by no means striving to extend its own remit by this means.

In addition, the industrialised nations in particular, which are primarily responsible for climate change, reject the term environmental refugee. Both UN organisations and representatives of industrialised nations constantly refer to the fact that, given the multifaceted and overlapping causes of migration, it is almost impossible to identify the impact of climate change as a main trigger of migratory movements, voluntary or otherwise, with the result that it cannot be proved that any flight is caused primarily by the effects of climate change.

Two scientists working on the EACH-FOR project, Olivia Dun and François Gemenne, counter this argument by pointing out that under the Geneva convention refugees are not anyway required to demonstrate persecution as the main reason for their migration, but rather, the decisive factor for granting refugee status is whether persecution in accordance with Article 1 has actually taken place or not. As soon as any association has been shown between persecution and flight, then according to Dun und Gemenne decision-makers could grant refugee status.

The Norwegian Refugee Council (NRC), which can identify no conclusive definition of the required state of persecution in the UNHCR regulations, also believes that it is entirely possible to recognise climate change as a form of persecution. Thus Paragraph 53 of the *UNHCR Handbook on Procedures and Criteria for Determining Refugee Status* provides for the recognition of refugee status on the basis of "cumulative grounds", not in themselves amounting to persecution, but which, if taken together "produce an effect on the mind of the applicant that can reasonably justify a claim to well-founded fear of persecution". According to the NRC, this concept leaves room for interpretation such that environmental refugees can be protected under the GRC and associated UNHCR regulations.

Moreover, human rights organisations assert that people affected by environmental migration are being robbed of their fundamental right to protection in a situation similar to that of refugees. These people are, by virtue of this, permanent refugees and should therefore also be treated as such. A corresponding category of *environmental refugee* is therefore only logical. Moreover, the migratory movement is a reaction to an externally induced circumstance, similar to a threat or persecution as provided for by the GRC as a condition of refugee status. The organisations therefore plead both for the introduction of the term *environmental refugees* and for an extension to the content of the GRC to recognise such people as "genuine" refugees.

The protection offered to environmental migrants is currently precarious. To date there is still no internationally recognised document requiring that the international community of nations should provide support for environmental migrants in the event that their country of origin is unable to do so. Existing regulations do not oblige international states to take in environmental migrants. Those agreements that do exist can either only be applied in exceptional cases or can be interpreted too broadly to offer reliable protection, or else they are only "can" regulations with no binding effect.

The effects of climate change, alongside other socioeconomic factors, are a trigger for existing and future migratory movements. In practice, however, it will be difficult to make a clear distinction between these triggers in order to identify environmental migration as demanded by some scientists. Specialist literature is divided on the subject of environmental migration. Whereas some scientists deny its existence and speak instead of economic and poverty-driven migration, others regard climate change to be the main reason for migratory movements worldwide.

Environmental migration, like every other social process, takes place within a socio-economic context, so that attempts to draw a precise dividing line between it and other causes of migration, such as war, poverty or climate change, are, in the author's view, doomed to failure from the outset. Nonetheless, it can be assumed that a considerable number of people will be confronted in coming decades with such phenomena as rising seas levels, expanding desert regions and a lack of fresh water. As a result, many of these people will migrate within national borders or across international borders either voluntarily or in flight. Nonetheless, economic, political and cultural aspects of migration must also be considered in order to take account of the complexity of environmental migration. It takes place under the influence of various push- and pull-factors so that answers based on a single cause are not sufficient.

It will be a great challenge in future to decide what status – and consequently what legal status – the affected people are to be granted. International legal norms provide too little protection for environmental migrants, partly due to the absence of any recognition of this new migration phenomenon. The Geneva Refugee Convention (GRC) and its additional protocols only consider some environmental migrants under certain circumstances and therefore do not offer any comprehensive protection. Only a few of today's environmental migrants satisfy the conditions of the GRC, so the majority of persons affected are not currently treated as refugees under current legal conditions. Nor do the legal instruments of nation-states or regions provide environmental migrants with comprehensive protection. It is therefore urgently necessary that regulations should recognise the phenomenon of environmental migration and be adapted to accommodate it. In order not to endanger existing categories, an additional protocol or a new convention appears more meaningful and likely of success than amending the

GRC. Furthermore, new regional and national agreements could additionally protect the rights of environmental migrants.

Since the responsibility for climate change rests primarily with the western industrial nations, they are especially responsible for those suffering environmental migration. How far they are ready to meet that responsibility – whether through taking in such people or by providing considerable support in lessening the impact of climate change – will be decisive for the protection of environmental migrants. However, the countries from which environmental migrants originate also have great responsibility towards their citizens and are obliged to do their best to protect their lives. They must take preventive measures to adapt to the consequences of climate change and lessen their impact over both the short and long term.

Climate change presents the international community with great challenges, which can only be overcome if communities work together. Dealing with environmental migration is one of those challenges. If appropriate measures are to be taken, then it is vital to gather additional information about environmental migration. Research into this area should therefore be significantly intensified.

MIGRATION IN INDIA

India as a nation has seen a high migration rate in recent years. Over 98 million people migrated from one place to another in 1990s, the highest for any decade since independence according to the 2001 census details. However in 1970s migration was slowing down. The number of migrants during 1991-2001 increased by about 22% over the previous decade an increase since 1951. Apart from women migrating due to marriage, employment is the biggest reason for migration. The number of job seekers among all migrants has increased by 45% over the previous decade. Nearly 14 million people migrated from their place of birth in search of jobs. The overwhelming majority of these-12 million was men.

Migrants have created pressure on others who are in same job market. While freedom to migrate within the country is an enshrined right the uneven development, levels of desperation and other factors have created friction points. Most people migrate because of a combination of push and pull factors. Lack of rural employment, fragmentation of land holdings and declining public investment in agriculture create a crisis for rural Indians. Urban areas and some rural areas with industrial development or high agricultural production offer better prospects for jobs or self-employment.

Contrary to common perception the search for jobs is more often within the same state than in some other state. About 9 million persons were intra-state migrants often within the district while 5 million went to other states. The intra-state figures include people moving from villages to nearby towns and cities in search of better jobs. Over 5.7 million persons who moved in search of jobs migrated from rural to urban areas. Another 4.5 million migrated within the rural areas looking for work.

The data shows that among people migrating in search of jobs, literates constitute the vast bulk over 10.6 million while illiterate migrants are about 3.3 million. Three out of four job-seeking migrants are educated males. Among literate, migrant job-seekers less than 1% was women. Nearly 40% of literate persons migrating for work had studied up to secondary level and another 32% had studied beyond. Graduates numbered over 1.8 million or about 17% while technical diploma or degree holders constituted about 8%.

About 72% do get regular work but over 11 million get less than 183 days of work in a year. This is a higher proportion than non-migrants. Independent NSS data from 1999-2000 indicates that migrant workers take up regular or casual employment or self-employment in nearly equal proportions. Around 8.1 million of the migrants were reported as available for or seeking work. The census data may not fully reflect seasonal or circulatory migration, estimated to be up to 10 million by the National Commission on Rural Labor. Seasonal migrants are usually dalits and other highly impoverished sections that go out to work in harvesting seasons or on construction sites, in brick kilns, salt mines etc. They go out to pay their debts and to survive.

Bibliography

Allen, P.: *Food For the Future: Conditions and Contradictions of Sustainability*, New York: John Wiley and Sons, 1993.

Aukland, L.: *The Effect of Selective Logging on Amphibian Diversity of Budongo Forest*, Uganda. B. A., Oxford University, 1997.

Bahati, J.: *Impact of Arboricidal Treatments on the Natural Regeneration of Species Composition in Budongo Forest*, Uganda. M. Sc., Makerere University, 1997.

Barlow K: *Ecology of Food and Nutrition*, Oceania, New Guinea, 1984.

Barrows, M.: *A Survey of the Intestinal Parasites of the Primates in Budongo Forest*, Uganda. Glasgow University, 1996.

Bauer, Steffan: *Handbook of Globalization and the Environment*, CRC Press, Delhi, 2007.

Bethell, E.: *Vigilance in Foraging Chimpanzees*. University College, London University, 1998.

Biswas, A., and Cline, S.: *Global Change: Impacts on Water and Food Security*, Springer, Heidelberg, 2010.

Burdak, L.R.: *Recent Advances in Desert Afforestation*, F.R.I., Dehra dun, 1982.

Burroughs, W.J.: *The Climate Revealed*, Cambridge University Press, Cambridge, 1999.

Carley, M., P. Jenkins : *Urban Development and Civil Society, the Role of Communities in Sustainable Cities*, Sterling, VA, Earthscan Publications, 2001.

Casado, Matt A: *Food and Beverage Service Manual*. New York: Wiley, 1994.

Chouhan, T.S.: *Desertification in the World and Its Control*, Scientific Publishers, Delhi, 1992.

Daubenmire, F.: *Plants and Environment*, New York, Wiley, 1947.

Geist, Helmut: *The Causes and Progression of Desertification*, Ashgate Publishing, Delhi, 2005.

Goodman, D.C.: *From Farming to Biotechnology: A Theory of Agro-industrial Development*, Oxford, Blackwell, 1987.

Katz, P., V. J. Scully : *The New Urbanism, Toward an Architecture of Community*, New York, McGraw-Hill, 1994.

Magalhaes, A. R. *Impacts of Climatic Variations and Sustainable Development in Semi-arid Regions*. Proceedings of International Conference. ICID. Fortaleza, Brazil, 1992.

Oliver, John E.: *Encyclopedia of World Climatology*, Springer, Delhi, 2005.

Perrings, C. : *Sustainable Development and Poverty Alleviation in Sub-Saharan Africa*, Botswana, New York, St. Martin's Press, 1996.

Pollock NJ: *The Concept of Food in Pacific Society: A Fijian Example*, Oceania, Fiji, 1985.

Reynolds, F.: *Chimpanzees of the Budongo Forest*, New York, Rinehart and Winston, 1965.

Shiel, D.: *The Ecology of Long-term Change in a Ugandan Rainforest*. D. Phil., Oxford University, 1997.

Simonds, J. O. : *Garden Cities 21, Creating a Livable Urban Environment*, New York, McGraw-Hill, 1994.

Sobsey M (2002) *Managing food in the home: Accelerated health gains from improved water supply*. Geneva, World Health Organization.

Tucker, Mary Evelyn, and John Grim: *Worldviews and Ecology: Religion, Philosophy, and the Environment*. Orbis Books, Maryknoll, N.Y. 1994.

Uhlir, P. F. : *Scientific Data for Decision Making Toward Sustainable Development*, Washington, D.C., National Academies Press, 2003.

Walter, B., L. Arkin : *Sustainable Cities, Concepts and Strategies for Eco-city Development*, Los Angeles, CA, EHM Eco-Home Media, 1992.

Webb, R. : *Floods, Droughts and Climate Change*, University of Arizona Press, NY, 2001.

Index

N

P

R

S

❑❑❑